TEACHING EFFECTIVELY WITH ZOOM

First edition. July 1, 2020

ISBN

Ebook: 978-1-7353408-0-7

Paperback: 978-1-7353408-1-4

The COVID pandemic has forced millions of teachers in elementary schools, high schools, colleges, and universities around the world to re-think their basic craft. How can we facilitate learning through a thin stream of electrons that produces fuzzy images and delayed audio? Every single one of us should read *Teaching Effectively with Zoom*. Written by one of the most effective teachers in the business, Dan Levy offers highly practical guidance about how to utilize the many features of the ubiquitous Zoom platform — from polling to chats to breakout rooms — to achieve a wide range of pedagogical objectives. This book will certainly make you a better teacher on Zoom, but it will also make you a better teacher in the classroom when we don't have to rely on Zoom anymore.

Archon Fung
Winthrop Laflin McCormack Professor and former Academic Dean at the Harvard Kennedy School
Harvard University

Thank you, Dan Levy, for taking us beyond backgrounds and breakout groups — and for putting the student experience at the center of the online learning discussion. The sudden COVID-induced shift to e-learning has been a ferocious jolt; in a blink, the methods used by K-12 teachers to assess student understanding in real-time, get students collaborating on rich and meaningful problems, and keep the learning exciting and engaging have been ripped away. In his guide, Levy combines a strong base of the science of learning with real-life examples from his classroom and those of his colleagues to put those learner-centric goals within reach again. And the techniques are doable! As I read, I kept thinking, "I can do that." Check out chapter 4 on the variety of ways to use polls to get a quick check on student understanding. In my career as a K-12 classroom teacher, instructional leader, and teacher developer, I've always wondered about how much attention higher ed professors pay to teaching outcomes. In 15 years of "talking shop" with Dan, I can say with great confidence that he's an outcomes geek, and I'm grateful that he's directed his energy toward this absolutely essential guide. Read it today, and reach more students tomorrow!

Dave Meyers
CEO and Co-founder of Teachers Connect

This eminently practical, reader-friendly guide offers instructional wisdom that will make newcomers' entry into teaching online easier, and gives even seasoned instructors some fresh strategies and tips. As Dan Levy pulls back the curtain on his own teaching journey, he shares a wealth of detailed explanations, frameworks, and grounded practice. His approach to the book mirrors his commitments as an educator and colleague: when learning comes first, effective teaching will follow.

Allison Pingree
Senior Instructional Coach, Harvard University

To my family, for your love and support.

To my students, for constantly inspiring me to become a better teacher.

Foreword

As you come across this title, you may be asking yourself: "Who exactly is Dan Levy, and why is an economist who teaches statistics at Harvard qualified to write a book about online teaching? Is it a book worth reading?" Those are fair questions. And I'm going to do my best to answer them.

I first got to know Dan in 2009, when he allowed me to audit his Quantitative Methods and Empirical Analysis class to fill a gap in my statistical knowledge. In the first class session, I was excited by the content but blown away by the teaching. An elegant lesson plan. Students leaning in to participate. Questions that invited deep reflection. Technology used to engage and make thinking visible. A seriousness of purpose, alongside humor and joy. It was something to behold.

Dan caught me at the end of class, and knowing that I studied teaching, asked me if I had any feedback. I shared some things that inspired me and pushed myself to come up with some elements that might be tweaked or changed. He asked me if I'd be willing to provide this feedback every class, and I agreed. The statistics quickly became secondary as Dan and I engaged in a semester-long dialogue about excellent instruction. And we have been talking about teaching ever since, through my time then as an instructional coach at the Harvard Kennedy School (HKS) and now at the Harvard Graduate School of Education (HGSE).

One way to define and identify teaching excellence is to look at the product. How well-crafted is the syllabus? How do students and peers evaluate the quality of the teaching? How much do the students learn? Dan excels in this realm. I use his syllabus as one of my exemplars when working with new faculty. Dan's students consistently rate his course off the charts. In a required statistics course with 60 students, where both "required" and "statistics" are elements not in the instructor's favor, Dan once had every single student give him the top rating (on a five-point scale) for overall quality of instruction – something I have not seen before or since. He has won school- and university-wide teaching awards, and is a featured faculty in Instructional Moves, a project started at Harvard to help educators all over the world to incorporate and refine high-leverage teaching practices. He has been elected to chair high-profile pedagogical and programmatic initiatives at HKS, and has conducted workshops on instruction throughout the U.S. and the world.

Another framing of teacher excellence is as a process, one of continuous improvement. In its essence, it's a process of identifying goals for you and your students, measuring how well the goals were met, and then tirelessly working to close the gap between the aspirational and the actual. I can honestly say that I have never met an educator more committed to continuous improvement than Dan. He analyzes student performance, combs through student evaluations, solicits additional student feedback, and reads books on pedagogy. He engages in extensive conversations with colleagues around teaching, pays numerous visits to classrooms across the university, and has an open-door policy for peers to observe him. When grappling with a particularly important pedagogical challenge, Dan turns his scholarly approach to his own teaching. He has conducted and published studies about the best way to get students to prepare for class, how to elicit student opinions through different polling mechanisms, and how to make exams themselves into significant learning experiences instead of just a means to measure learning. He then applies these new understandings into the next iterations of his teaching in the service of his students' learning.

Dan also applies this approach to teaching with technology. He has worked to incorporate hybrid elements into his in-person classes. When he couldn't find the right tools to get to know his students and their participation patterns better, he co-founded Teachly, a web application aimed at helping faculty members teach more effectively and inclusively. He volunteered to serve as the Faculty Co-Chair of the HarvardX Research Committee, which focused on promoting research about online learning, and he currently serves as the Faculty Director of the Harvard Kennedy School's flagship initiative on online education.

Then the global pandemic hit, turning the world and the world of education upside down. Dan was ready to respond.

He began to work through the problems that arose: Do I need to modify my goals and my assessments? How should I balance synchronous and asynchronous approaches? What tools should I use? How do I create experiences that build a sense of community and unity when people are by themselves on their own devices? How do I create the conditions for quality engagement and intellectual rigor? How do I attend to issues of access and equity for all of my students?

Dan consulted the existing literature. Although he found a lot of good ideas, he also found some notable gaps, particularly about how best to leverage synchronous teaching tools like Zoom to create effective learning experiences. These gaps are understandable. The online tools at our disposal and people's facility in using them continue to evolve at lightning speeds.

So Dan dove deeper. He tried to digest and make sense of the deluge of advice columns and tip sheets that emerge on a daily basis. He consulted online learning experts across campus and beyond. He brainstormed new ideas and bounced them off colleagues. He experimented and gathered feedback on the approaches. He interviewed students to get their perspective on the various approaches their instructors were taking. Eventually, when he and a colleague transformed a highly rated executive education program into an exceedingly well-received online offering in short order, the idea to write this book was born. As he said to me: "A book that can be a one-stop shop for busy instructors who want to teach well using something like Zoom needs to be out there, and I couldn't find it. So I feel like I should be of service and do my best to write it."

The fruits of that commitment, and all the labor that followed, resulted in this guide, which embodies Dan's approach to teaching. Focus first and foremost on the learning. Start with your goals and work backwards. Attend just as carefully to the details as to the big picture. Treat teaching as a profession by helping instructors expand the tools at their disposal and the discernment of when and how to use them. Be thoughtful about variations in context and their implication. Do everything you can to have students feel engaged, included, and empowered.

As someone dedicated to helping higher education instructors improve their practice, I am thrilled to have this book to share with my colleagues. As someone who previously taught in middle and high school and who is married to a middle school instructional coach, I am similarly excited about how the ideas in this book can add value to PK-12 educators. The gift I first had as an observer and thought partner in Dan's statistics class – sharing in his insights, his curiosity, and his humanity – is now available to you. I hope and trust you will benefit from it as much as I have.

Josh Bookin
Associate Director, Instructional Support and Development
Harvard Graduate School of Education

Preface

In early 2020, because of COVID-19, many colleges and schools around the world closed, and many teachers, instructors, and faculty members had to learn how to teach online in a hurry. Since then, a lot has been written about how we instructors should teach live online sessions and how we should master the different aspects of the technology involved. How do we set up windows on our screens when we teach online? Should we use the chat function? Do we use breakout rooms, and if so, how? For many of us, it was overwhelming. Even paralyzing.

This book is an attempt to help educators take a step back and make sense of all the chaos. My hope is to provide you with some useful pedagogic principles and practices that have served me and some of my colleagues well, and that I hope will guide you in your journey to develop your online teaching skills. To help ensure that you can immediately apply the ideas in this book, I have included screenshots to illustrate how to implement some key practices in Zoom. But because the technology is evolving quickly, some of the instructions on how to implement these practices (including links to videos and screencasts) are described in the companion site (https://www.teachingeffectivelywithzoom.com/), which can be updated more quickly than the book. The companion site also contains links to additional resources, including readings and checklists.

Whether you are new to online learning or a veteran, I hope you will find something of value in the book. If you are new to teaching online, my hope is that you will find a few ideas, try them out, and then experiment with additional ideas later in your journey. I also hope that this book will help you focus on what is important and provide a roadmap in your efforts to learn how to teach online. Hint: Spending 15 hours arranging monitors on your desk, wondering how to set up the various Zoom windows, and experimenting where to put your microphone is not an efficient use of your time. I learned this the hard way! If you are an online teaching veteran, I hope that some of the meta-advice here will be helpful and that you will pick up a few tips to improve your already well-developed practices.

The book is based on my own experience teaching online, observations of several colleagues teaching online, research-based principles of effective teaching and learning, and, perhaps just as importantly, interviews with dozens of students who recently experienced online learning for the first time and also had to adapt to this way of learning in a hurry.

Who is this book for? I teach at a university, so one natural audience for this book is people with teaching responsibilities in colleges and universities (faculty, instructors, instructional coaches, teaching assistants, etc.). But after observing my daughters go through more than a handful of classes over Zoom in these past few weeks, I am convinced that K-12 teachers can also find value in this book.

The book's main focus is how to conduct effective **live** classes online. I am aware that a live class session – the activity you and your students synchronously engage in together – is only part of the larger ecosystem of teaching online that typically includes asynchronous assignments and activities before and after class. And I am a big believer that the best online teaching happens when we combine synchronous and asynchronous approaches, and use each of these to their comparative advantages (see chapter 10). But there are already a lot of good general references for online teaching and my sense is that what is most needed at the moment is guidance on how to effectively design and conduct *live online* sessions that are compelling learning experiences for our students.[1]

Why Zoom? It's the predominant platform right now. It will make the advice in this book more concrete and grounded. Nevertheless, a lot of the advice here applies to many other platforms used to teach live sessions, such as Webex, BigBlueBotton, Google Hangouts, Microsoft Teams, and others.

The book is organized as follows. **Part I** provides an introduction and a brief overview of the *key overarching principles* that I hope can guide your online teaching journey. **Part II** reviews ways in which *students can engage* in a live online class (speak, vote, write, work in groups, and share their work) and the corresponding Zoom tool(s) involved. **Part III** reviews ways in which *you can engage* when you teach a live session and the corresponding Zoom tool(s) available to you. It focuses on presenting (slides or other material) and on annotating. The choice of addressing ways students can engage before ways you can engage is a deliberate one. It reflects an underlying belief of this book, and my own teaching, that the student comes first. It also helps combat our natural tendencies as teachers of spending most of our time thinking about what we will be doing in class rather than what our students will be doing. **Part IV** is meant to help put some of the things you learned in a broader context and to pull things together.

Over the past few years, I have read a few books on teaching and learning that have influenced the way I teach. I would like to mention two in particular: Derek Bruff's *Teaching with Classroom Response Systems* (2009) and Jim Lang's *Small Teaching* (2016). While I was reading these books, my brain was constantly coming up with ways of applying, adapting, or tweaking some of their ideas to my own teaching. It was exhilarating. My hope is that your experience reading this book will be similar. I hope you will come up with many ideas to implement in your online teaching.

If this book inspires you to do something you find interesting or exciting when teaching online that translates into your students engaging and learning more, please send me a note telling me about it. Nothing would be more rewarding for me.

Dan Levy
Cambridge, MA.
June 2020

Acknowledgments

I would like to acknowledge the help of many people who have made this book possible. Five people deserve special mention. First, my friend, colleague, and Harvard Business School professor Mike Toffel, who gave me insightful feedback on every chapter of the book, helped refine my thinking, and supported me in numerous ways throughout the writing process. Second, my student Ruth Hütte, who committed wholeheartedly to this project and contributed to the book in many ways, including figuring out the process of publishing to make the book accessible to as many educators as possible. Third, Yamile Nesrala was an excellent and careful editor, and committed to being a partner on the project beyond what a regular editor does. Fourth, Teddy Svoronos, a colleague and friend whose use of technology to teach effectively is second to none, and who kindly and patiently answered many questions while I was writing this book. Last but not least, Victoria Barnum, who led the effort to produce the companion site and helped me have the time needed for writing.

I would like to thank all the educators that allowed me to learn from their online teaching practices by observing them teach and/or speaking with them, including Carrie Conaway, David Eaves, Terry Fisher, Marshall Ganz, Jim Honan, Dutch Leonard, Zoe Marks, Rebecca Nesson, Rem Koning, Shoshanna Kostant, Kathy Pham, Allison Shapira, Mark Shepard, Teddy Svoronos, Mitch Weiss, and Julie Wilson. Some of them are featured in the "In Practice" sections of the book.

I am also grateful to many colleagues who have influenced my teaching perspectives, including but not limited to Alberto Abadie, Bharat Anand, Matt Andrews, Arthur Applbaum, Chris Avery, Mary Jo Bane, Matt Baum, Erin Baumann, Bob Behn, Joe Blitzstein, Iris Bohnet, Derek Bok, Peter Bol, Josh Bookin, Jonathan Borck, Dana Born, Matt Bunn, Sebastian Bustos, Filipe Campante, Gonzalo Chavez, Suzanne Cooper, Anjani Datla, Jorrit de Jong, Akash Deep, Pinar Dogan, Jack Donahue, Susan Dynarski, Erin Driver-Linn, Greg Duncan, David Ellwood, Doug Elmendorf, Mark Fagan, Carol Finney, Maria Flanagan, Jeff Frankel, John Friedman, Archon Fung, Alan Garber, Patricia Garcia-Rios, Steve Goldsmith, Tony Gomez-Ibañez, Josh Goodman, Merilee Grindle, John Haigh, Sarah Hamma, Rema Hanna, Frank Hartmann, Ricardo Hausmann, Ron Heifetz, Dave Hirsh, Andrew Ho, Daniel Hojman, Jim Honan, Kessely Hong, Deborah Hughes Hallett, Anders Jensen, Doug Johnson, Tom Kane, Felipe Kast, Steve Kelman, Alex Keyssar, Adnan Khan, Asim Khwaja, David King, Gary King, Mae Klinger, Steve Kosack, Maciej Kotowski, Michael Kremer, Robert Lawrence, Henry Lee, Dutch Leonard, Jennifer Lerner, Jeff Liebman, Dick Light, Rob Lue, Erzo Luttmer, David Malan, Brian Mandell, Jane Mansbridge, Tarek Masoud, Janina Matuszeski, Quinton Mayne, Eric Mazur, Tim McCarthy, David Meyers, Matt Miller, Nolan Miller, Francisco Monaldi, Mark Moore, José Ramón Morales, Juan Nagel, Angelica Natera, Tim O'Brien, Rohini Pande, Tom Patterson, Allison Pingree, Roger Porter, Samantha Power, Lant Pritchett, Todd Rakoff, Fernando Reimers, Hannah Riley-Bowles, Juan Riveros, Chris Robert, Chris

Robichaud, Dani Rodrik, Todd Rogers, Lori Rogers-Stokes, Eric Rosenbach, Jay Rosengard, Soroush Saghafian, Tony Saich, Miguel Angel Santos, Jeff Seglin, Kathryn Sikkink, Judy Singer, Malcolm Sparrow, Rob Stavins, Guy Stuart, Federico Sturzenegger, Arvind Subramanian, Karti Subramanian, Moshik Temkin, Dustin Tingley, Tamara Tiska, Rodrigo Wagner, Jim Waldo, Steve Walt, Michael Walton, Lee Warren, Rob Wilkinson, Julie Wilson, Carolyn Wood, Michael Woolcock, Josh Yardley, Andrés Zahler, David Zavaleta, Richard Zeckhauser, and Pete Zimmerman.

Many students provided me with valuable insights for the book, including but not limited to Diego Auvert, Nicole Carpentier, Roukaya El Houda, Jossie Fahsbender, Maria Fayos Herrera, David Franklin, Sophie Gardiner, Catri Greppi, Fatine Guedira, Varun Gupta, Anastacia Kay, Casey Kearney, Chris Kranzinger, Shiro Kuriwaki, Julia Liniado, Megan Linquiti, Zainab Raji, Eki Ramadhan, Alicia Sikiric, Racceb Taddesse, Jiawen Tang, Montse Trujillo, Beatriz Vasconcellos, Hannah Wang, and many who participated in focus groups and informal sessions I conducted. Their perspectives, insights, and candor tremendously shaped my views and thinking.

I am grateful to others who helped me with research, advice, insight, access to key information, and/or encouragement, including Manuel Alcalá, James Brockman, Piet Cohen, Kate Hamilton, Andy Levi, Horace Ling, Anne Margulies, Maddie Meister, David Meyers, Anna Shanley, Kristin Sullivan, and Ian Tosh.

Finally, I am very grateful to Josh Bookin, Allison Pingree, and Carolyn Wood for shaping who I am as a teacher, for supporting me in so many ways over the last decade, for helping me throughout the process of writing this book, and for not saying to me "this is a crazy idea" when I first told them that I wanted to write this book and that I had to do it in one month for it to be useful to the world.

Table of Contents

PART I – KEY IDEAS

This part provides the broad frame for the rest of the book. Chapter 1 introduces the book, explains what it is about, its key sources, its approach, and how it is organized. It ends with a section on how to get started with Zoom. Chapter 2 lays out the key pedagogic principles that underlie the teaching practices and a lot of the advice provided in this book.

Chapter 1 – Introduction

Most people reading this book did not *choose* to teach online. Most of us were forced to do so by Covid-19. I hope that by the end of this book, you will have gained skills to improve how you teach online, and perhaps even some appreciation for the things that you can do online better than in person. But I want to start this book by acknowledging the profound sense of loss that many of us as educators are experiencing as we transition from teaching in a physical classroom to doing so in a virtual one.

The experience of teaching a class in person can be profoundly human. Before class begins, we might chit chat with some students near us, warmly welcome others as they come in the room, notice a student who seems tired or sad and go and check in on them, have a student stop by and tell us something that happened to them over the weekend that reminded them of our class, and so on. As the students come in, they say hi to each other, sit down, and start speaking with the other students around them. It is not unusual to observe laughter and joy, and sometimes tears. You can feel the energy in the room; sometimes high, sometimes low, but you can feel it.

As class begins in a physical classroom, you might start presenting or explaining something, and perhaps take for granted that being with your students in the same physical space helps define both the spatial geography at your disposal and the boundaries of your learning community. You know that Maria, who sits at the back of the room, had a difficult week and you might try to engage her in a different manner today. As the class progresses, you might refer to an observation that Pat, who sits on your right, made a few minutes ago. And you can walk around the room and use verbal and non-verbal language to communicate with your students. You might approach John with a gentle smile, or a stern look (depending on your style), when you want him to stop doing whatever he is doing. John will notice this; he can see you are looking at him and not anyone else. As the class progresses, the students interact with you and with each other in natural and human ways. They might laugh at something funny, and everyone in the room hears that laugh. And yes, they might interrupt or disrupt the class in ways that you might not like. But it is a human experience, similar in many ways to the ones you might have when you get together with a group of people anywhere. There is no need for anyone to unmute themselves or click a button to spontaneously express their reactions to what is happening live in the classroom.

As class ends, you stay in the classroom for a little while longer, and might have some students come and ask you a question, make a comment, or simply say goodbye. As you leave the room, you might bump into a colleague or two and catch up with each other.

Despite the remarkable technological advances over the last decade, the experience of teaching a live online class at this time feels very different. The physical classroom seems to have been flattened onto a computer screen. If you have more than a handful of students, they typically mute themselves and only unmute when they talk to the class. Dialogue feels less natural. You cannot hear the spontaneous laughter in the room. And despite your best intentions, John cannot feel that you are looking at him and only him.

So while many educators and education leaders might rightly encourage us to see the opportunities brought about by online learning, I think it is hard to do this without first acknowledging the loss.

Another loss that it is important to recognize is the one that our students are experiencing. Again, many of them did not choose to learn online. And while they may be more digitally sophisticated than many of us, they too will miss many of the things described above and more. They will miss the interactions with each other, and the spontaneous connections that are so integral to a well-rounded education.

So my first piece of advice before you continue reading this book is to pause for a moment to reflect on these losses and what they mean to you in particular. Doing so will allow you to clarify the values that you want to advance in your online teaching, and will put you in a better position to embrace its challenges. It will also make your reading of this book more productive and enjoyable.

What this book is about

This book is about helping you teach effective live online sessions with your students using Zoom. Whether you chose to teach online or were thrust into it due to school building closures, I hope this book will provide you with some guidance and ideas on how to teach effectively remotely. There is so much advice out there that it can be overwhelming. And on top of it, there is an increasing amount of technology to master. Part of my goal in writing this book is to help you focus on what's important. You have limited time. You cannot spend countless hours aimlessly googling how to teach on Zoom. I tried to keep this reality in mind as I wrote the book.

The book is not a comprehensive guide to everything you can do on Zoom to teach an online live session, but rather a guide on where to focus your attention and limited time. The overarching theme is that at the end of the day, you are there to help your students learn. Period. It is that simple. The technology is just the vehicle. You need to master some of the technology so you can focus on that goal. But you don't need to become a Zoom master to do this well.

The book focuses on some useful pedagogic principles and practices that have served me and some of my colleagues well, and that I hope will guide you in your journey to develop your online teaching skills. I use the term "develop your online teaching skills" deliberately. I would like to nudge you to think of this as a process where you will get better with preparation and practice. Just like you have done with other aspects of your job.

In sum, my goals in writing this book are to give you some ideas, empower you, and ultimately inspire you to teach effectively with Zoom. My goal is not to overwhelm you or try to convert you into a Zoom expert.

What this book is not about

This book is not about online learning in general. I recognize that live classes are only one part of the larger ecosystem of teaching online. And I am a big believer that the best online teaching happens when we combine synchronous and asynchronous approaches and use each of these to their comparative advantages. In fact, chapter 10 is devoted to how you can integrate online live sessions with other material you provide (e.g., online modules, videos, quizzes, readings, etc.) for students to engage with on their own time. But there are many good general references for teaching online,[1] and my sense is that what is most needed at this moment is guidance on how to effectively design and conduct *live online* sessions that are compelling learning experiences for our students.

What this book is based on

The book is based on my own experience teaching online, observations of several colleagues teaching online, research-based principles of effective teaching and learning, and, perhaps just as importantly, interviews with dozens of students who recently experienced online learning for the first time and also had to adapt to this way of learning in a hurry.

Although many of the approaches and practices recommended in the book are backed by research-based principles of effective teaching and learning, the focus will not be on the exposition of these principles and research findings. If you are interested in them, I recommend exploring the sources listed at the end of the book that pique your interest. The following books might also be of use to you if you are interested in learning more about some of the underlying research base of the learning sciences:

- How People Learn: Brain, Mind, Experience, and School (2000) by the National Research Council

- *How Learning Works: 7 Research-Based Principles for Smart Teaching* (2010) by Susan A. Ambrose, Michael W. Bridges, Michele DiPietro, Marsha C. Lovett, and Marie K. Norman

- *Make it Stick: The Science of Successful Learning* (2014) by Peter C. Brown, Henry L. Roediger, III, and Mark A. McDaniel

- *Multimedia Learning* (2009) by Richard E. Mayer

What is the approach of this book?

This book is designed to be practical. I want you to gain some concrete ideas to use in your live online classes. I have included screenshots to illustrate how to implement some key practices in Zoom. But because the technology is still evolving quickly, some of the instructions on how to implement these practices are described in the companion site, which can be updated more quickly than the book. I also want to invite you use the companion site to share your own practices and see what other readers have shared. I hope the next edition of this book will feature practices from the readers in the "In Practice" sections (see below).

How is this book organized?

To help ground your learning, the following sections appear in most chapters of the book:

Table 1.1 – Chapter sections

Title		Goal
In Practice		Describes how a real educator has used an innovative practice to teach effectively with Zoom.
Checklist		Provides a sample checklist of the concrete steps you need in order to implement the ideas of that chapter. Editable versions of the checklists are on the companion site to enable you to customize these.

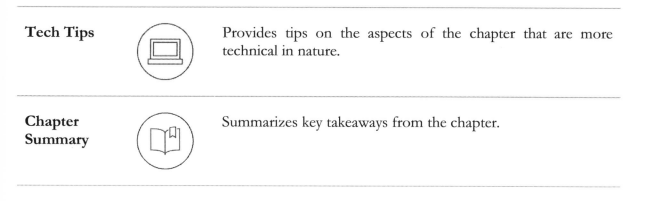

| **Tech Tips** | | Provides tips on the aspects of the chapter that are more technical in nature. |
| **Chapter Summary** | | Summarizes key takeaways from the chapter. |

Context matters

I have tried to write this book to help educators in all sorts of institutions teach better with Zoom. In doing so, I have sought to recognize differences between institutional contexts. Some of you will have IT departments to consult with, teaching coaches to guide you, resources to meet digital accessibility guidelines, and teaching assistants to help you. Others will just have a Zoom account and be pretty much on your own in figuring things out. Moreover, some of you can expect all or most of your students to be able to connect to your online live classes via video using a laptop from a quiet place, whereas others will be teaching students who might be able to connect intermittently by phone. While I have tried to write the book in a way that, whatever your circumstances, you can get valuable ideas for your teaching, I recognize that context matters and that you will likely need to tweak and adapt some of these ideas to your institutional setting. My goal is for you to find *some* ideas useful to improve your teaching.

How to read this book

The book is meant to be read in order, as there is a logical progression. But the chapters are fairly self-contained, so if you are particularly interested in a chapter, you can skip ahead. My advice is that regardless of the order in which you read it, you keep a notebook (in paper or electronic) near you so you can write the ideas from the book that you want to try out, and perhaps some ideas not in the book that occurred to you while you were reading the book, that you also want to try out. Otherwise, it is too easy to forget these ideas. After you are done reading the book, you can review these ideas and see which ones you most want to pursue. If you are new to online learning, you might want to re-read (or skim) some sections of this book after teaching for a few weeks, as you will likely have a new perspective and be ready to try some new approaches.

Getting started with Zoom

The book will assume that you have some familiarity with the basic Zoom interface from having participated in some Zoom meetings. In particular, it is important that you be comfortable with what Zoom calls the "Host controls," which are located on what I will refer to as Zoom's main toolbar (see below), and that you know how to set up a meeting in Zoom so that others can join. If you are unfamiliar with these aspects of Zoom, please check some introductory videos on the book's companion site.

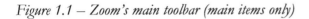

Figure 1.1 – Zoom's main toolbar (main items only)

<u>Tweaking Zoom settings</u>

Zoom has many settings that you can change to make some features available or to change the defaults in some features. Which features are available and which defaults are in place might depend on the way that your institution has set things up. To make full use of the advice presented in this book, it will sometimes be necessary for you to change your Zoom settings so that some features become available in your Zoom main toolbar and/or so that some defaults are consistent with your preferences. The "Tech Tips" section in several chapters provides guidance on how to make these changes, or will refer you to the companion site for further instructions. Below is one example of a Zoom toolbar with some additional features including polling and breakout rooms, which will be discussed in chapters 4 and 6, respectively. Your current Zoom bar will not necessarily look identical to this.

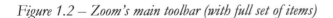

Figure 1.2 – Zoom's main toolbar (with full set of items)

<u>Organizing Zoom windows</u>

Zoom has many windows and possible views, which is great in providing you with a lot of information useful to teach a live session, but can sometimes become hard to manage. Below are two typical views showing your students, the participants list, the chat window, and Zoom's main toolbar. The first view shows your students in Gallery View (which maximizes the number of students you can see at once), whereas the second one displays Speaker View (which has the student speaking in larger view). The exact view that you prefer will depend on your personal preferences, the size and number of monitors, the features most important to you, and the activity you are doing at that moment (e.g., sharing slides, leading a discussion, etc.). I think the only way to figure out your optimal setups is by experimenting and practicing, but being able to see many of your students while you are teaching is probably a key part of any setup.

Figure 1.3 – Zoom's windows arrangement under "Gallery View"

Note: Students consented to appear in this picture.

Figure 1.4 – Zoom's windows arrangement under "Speaker View"

Note: Students consented to appear in this picture.

The maximum number of student thumbnails Zoom can display in a single screen depends on your computer's processor. If you exceed this number, Zoom will show you left and right arrows to navigate to additional pages where you can see additional students.

Equipment Needed

A natural question you might be asking is what kind of equipment you need to teach effectively on Zoom. The answer will depend on your preferences and resources available, but here is a list of equipment that would be useful, along with my sense of how important each item is.

Table 1.2 – Equipment needed

Equipment		Notes	Importance
A computer		Ideal device is a desktop or laptop computer. Pure tablets (like iPads) are good as an annotating device but not ideal for driving your Zoom class.	Very high
A good internet connection		Ideally an ethernet connection. If not, a reasonably fast and stable wi-fi connection.	Very high
A device to annotate what students say (e.g., blackboard, flipchart, document camera, tablet)		For pedagogic reasons, this is helpful. See chapter 9 for details and to help you make the choice.	High
Headphones		These are important so you can listen as clearly as possible. Might be combined with a microphone if needed (see below).	High
A microphone		How much you need this depends on how good your current one is.	Medium

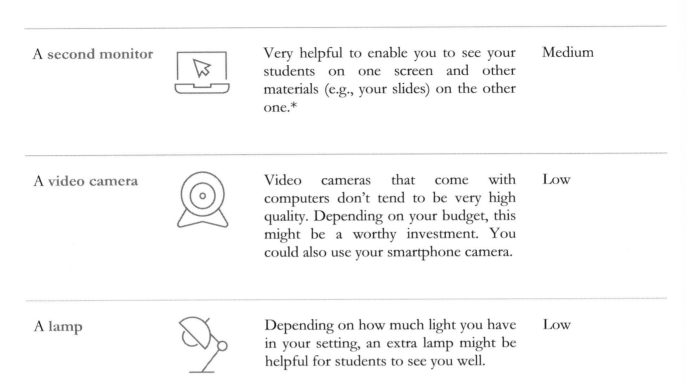

A second monitor		Very helpful to enable you to see your students on one screen and other materials (e.g., your slides) on the other one.*	Medium
A video camera		Video cameras that come with computers don't tend to be very high quality. Depending on your budget, this might be a worthy investment. You could also use your smartphone camera.	Low
A lamp		Depending on how much light you have in your setting, an extra lamp might be helpful for students to see you well.	Low

*Zoom has a specific "dual monitor" mode that allows for gallery view, speaker view, shared screen, chat, and participants in different windows. See chapter 8 for details.

The companion site has links describing how some educators have set up their workspace to teach in Zoom.

Disclaimer

The Zoom platform is likely to evolve quickly over the next few years. This book was written based on Version: 5.0.2 (24030.0508) of Zoom for the Mac OS, the latest version available in mid-June 2020. It is based on the version of Zoom's product designed for conducting classes, not webinars.

While some of the images on the screenshots will look differently on a Windows computer, the functionality across Mac and Windows is almost identical. I have tried to write the book so that most of what is in it will remain true for some time. This is certainly true of the pedagogic advice. And I have relegated most of what I think will evolve over time to the companion site. But there are aspects of what's written here that will need to be updated in the future. So you can assume that every description of a Zoom feature or its default settings in this book should be preceded by "As of the time of writing, ..." I will be keeping close track of the changes to the platform that affect the advice in this book, and invite you to alert me to these changes in the feedback section of the companion site.

 **Tech Tips**

Zoom's features and default settings vary by what type of account you have and by your institution's default settings. As you discover features you like, I recommend that you tweak your default settings to conform to your preferences. As a reference, the table below has the current features and default settings related to teaching for my institution and for Zoom's basic (free) account. One limitation of the basic account is that the polling feature is not available.

Table 1.3 – Variations in Zoom features and default settings

Item	My institution	Basic
Audio Type	Telephone and Comp. Audio	Computer Audio
Join Before Host	On	Off
Co-Host	On	Off
Polling	On	N/A
Whiteboard – Autosave	On	Off

Nonverbal Feedback	On	Off
Breakout Room	On	Off
Closed Captioning	On	N/A
Waiting Room	Off	N/A
Allow Streaming of Meetings	On (FB, YouTube, Custom)	N/A

Note: I am very grateful to Ian Tosh for his help producing this table.

Chapter Summary

- Transitioning from a physical classroom to a virtual one represents a loss. Acknowledging this loss is important to be able to move forward with online teaching.

- Remember that you are there to help your students learn. The technology is just the vehicle.

- You need to master some of the technology so you can focus on helping your students learn. But you don't need to become a Zoom master to do this well.

- My main objective in writing this book is to give you some ideas, empower you, and ultimately inspire you to teach effectively with Zoom.

Chapter 2 – Guiding Principles

If you are transitioning from teaching in a physical classroom to a virtual one, it might be helpful for you to think about the pedagogic principles that underlie your teaching practices, and how best to apply or enact these principles in your live online classes.

To help you do so, this chapter focuses on five key principles that underlie a lot of the advice provided in this book:

- Principle 1 – Be student-centered.

- Principle 2 – Plan for active learning.

- Principle 3 – Begin with the end in mind.

- Principle 4 – Use online teaching to its comparative advantages.

- Principle 5 – Teachers are made, not born.

Maybe the principles seem so obvious that you will want to skip this chapter. Or maybe some of the advice contained within each principle might be inconsistent with your own principles, or perhaps with your teaching practices. Regardless, I hope this chapter will be helpful to you by making you aware of the underlying beliefs that drive this book, and by helping you put the practices contained later in the book in a broader framework that you can use to come up with your own online teaching practices.

In the rest of this chapter, I briefly describe each of these principles and how I think they make teaching and learning more engaging and effective.

2.1   Principle 1 – Be student-centered

Below is my favorite cartoon of all time:

Figure 2.1 – Teaching how to whistle

Source: Tiger © King Features Syndicate Inc.

The notion that you teaching something is not the same as your students learning it seems so obvious that you might think it does not even deserve to be mentioned. Well, I think that while most of us recognize this notion, our actions are not always consistent with it. When designing courses (or individual classes), we tend to focus more on figuring out what topics we will address than on what our students will be able to know/do/master as a result of their learning. In other words, we tend to focus more on what we will cover than on what our students will uncover. We tend to spend a lot of our class preparation time planning what *we* will do in the classroom (deciding how we will explain things, designing and tweaking PowerPoint slides to no end, etc.) and not as much time thinking about what *our students* will be doing in the classroom. When teaching a class session, we often start executing a class plan without really understanding what background, skills, and misconceptions our students bring that might affect their ability to learn what we are trying to teach them. Or we might rush through a lot of slides at the end of class when we are running out of time, thinking that doing so means we "covered" that material, even though we know that's not really so. And so on. If you have not fallen prey to some of these tendencies, hats off to you. But I think most of us have because as human beings we tend to spend more time on things that are under our control (the teaching) and less time on the things that are not (the learning of our students).

Being student-centered is not just about recognizing that teaching is not the same thing as learning. It is also about seeing teaching as a human enterprise where your connection to your students, their connection to you, and their connection to each other are at the core. This implies a deliberate effort to learn about students (their names, their backgrounds, etc.), to get to know them, and to try as best as you can to understand what is going on in their lives that might be enabling or impeding their learning.

I remember a conversation early in my career with Lee Warren, a legendary former teaching coach at Harvard University, where she was telling a group of faculty members that it was important for us to understand everything that could affect our class, whether students were tired or stressed, what else was happening in their program or at the university, the weather outside, etc. I thought this was absurd and told her, "I cannot control any of these things; I am there to teach the students!" Over the years, I have come to realize how wise her advice was. It is true that you cannot control these things, but they do affect the way your class will go, and learning to acknowledge and to adjust to these circumstances is, in my mind, part of what teaching is about.

During the spring of 2020, students at my university reported several challenges related to the transition to online learning, including getting easily distracted, difficulty staying motivated, and feeling socially isolated. While none of these challenges were under the direct control of the instructors, I think those who acknowledged these circumstances and tried to help students address these challenges were likely more effective at stimulating their students' learning. The particular challenges faced by your students might be different than these, but if you are interested, the companion site has a link to a presentation with these challenges and the solutions that students themselves proposed to overcome them.

In sum, empathy for your students is at the core of teaching. No amount of technology will change this. Technology is just a tool to help you in this endeavor. As Derek Bruff argued in his recent book *Intentional Tech,*[2] teaching and learning goals should drive instructors' technology use, not the other way around. And as Bharat Anand, Harvard's Vice-Provost for Advances in Learning, argued, "Ultimately, what's far more important in creating impactful learning experiences, whether in-person or online, are the principles of effective pedagogy. How do you spark curiosity, challenge assumptions, enable discovery, and inspire learning?"[3]

One of my concerns with our move to online learning is that in our efforts to master the technology, we forget that teaching is inherently human.

In Practice 2.1 – Focusing on what's important

Mitch Weiss teaches at the Harvard Business School and serves as the course head for a required course called "The Entrepreneurial Manager." His faculty team spent countless hours during spring break 2020 preparing to transition their course from in-person to online. After all the training, he wrote to his faculty team:

"We've spent so much time on the technology of all of this, that I wanted to remind us of the original technologies for teaching well in the case method: **curiosity** for what the students think and **empathy** for what they want to know. If you can still have ample stores of both, you will be fine. So, my advice is to do your best at not letting the screen get in the way of really listening and wondering, and thereby ultimately teaching."

<u>Applying this principle to teaching online</u>

- **Get to know your students.** Getting to know your students online is generally harder than getting to know your students in person. It will require more deliberate effort on your part. See chapter 11 for some ideas.

- **Build community.** Building community online is crucial and also requires more deliberate effort than doing so in person. See chapter 11 for some ideas.

- **Understand your students' circumstances.** Understanding the circumstances your students are facing in their online learning journey is crucial for designing learning experiences, establishing norms, and enforcing these norms. Find ways to connect with your students and learn about their circumstances, and try to adapt your approach to these circumstances.

2.2   Principle 2 – Plan for active learning

Years of research have shown that students don't learn much when their role is conceptualized as passive receivers of knowledge.[4] Instead, effective teaching requires creating meaningful learning experiences for our students that support their playing an active role in their learning, including opportunities for active processing, application, inquiry, and interaction with others and with the course material. "The one who does the work does the learning" is one compact way to encapsulate this principle.[5]

Yet despite this research and statements by many educational experts – including Nobel laureate Carl Wieman, who once argued that the college lecture is the educational equivalent of bloodletting[6] – urging us to employ teaching methods where our students are active learners, too many of us still employ teaching methods where students are not playing an active enough role in their own learning.

If your teaching approach is driven by active learning principles and practices, please feel free to skip ahead to the next principle. If you are not there, but are open to making some changes in your teaching, a great place to start is by reading Jim Lang's excellent book *Small Teaching*, where he suggests small changes you can make to incorporate principles from the learning sciences into your teaching. For an inspirational video on incorporating active learning methods into your teaching, I recommend watching "Confessions of a Converted Lecturer" by Harvard Physics professor Eric Mazur. Links to both of these resources are on the companion site. I recently asked a student how online learning was going for her, and she replied in frustration, "it's like watching YouTube videos all day." Whatever we do as educators, we must create a better learning experience for our students than this!

If you are not convinced about how ineffective it is to have students play a passive role in their learning, I suggest you do one thing: sit in a straight lecture and see what it feels like. You can do this online. I have three predictions. First, I think you will realize how hard it is to stay engaged and especially to remember what the speaker said an hour after the lecture. Second, I think that you will convince yourself that, in your classes, you don't speak as much as the lecturer you just saw. If this is the case, I urge you to measure in your next class the fraction of class time when you are speaking. You can do this by having someone time you with a stopwatch or by using technology (see companion site for details). My third prediction is that you will severely underestimate the percent of time you are speaking in class. I certainly did!

One common piece of advice for teaching online is that you should plan for interaction and that students should not stay in any activity for more than 5 minutes. While I applaud the effort to get us to employ more active learning approaches in our classes, I fear that this advice might be construed as "plan for hyperactivity" and might end up being less effective than it could be. The goal is not just to get students to be active, but to do so with a clear intention of what you are trying to achieve (see next principle).

- **Design for engagement in live classes.** For every online live class session you teach, think about how your students will engage (see part II of this book for a list of ways they can engage) and plan a variety of activities that will lead them to make meaning of the material, and ultimately achieve your learning objectives.

- **Design for engagement outside of your live classes.** When designing asynchronous materials (ones that students will engage with on their own time outside your live classes), think about actively engaging students with activities that require them to answer questions, do things, and/or reflect on their learning. See chapter 10 for more details.

2.3   Principle 3 – Begin with the end in mind

When I was a Ph.D. student at Northwestern University, I signed up to be a teaching assistant and attended a one-hour training session by the person who headed the teaching and learning center there. The fact that more than 20 years later I still remember several things I learned in this session is a testament to how effective it was. The person conducting the training was Ken Bain, who later went on to write *What the Best College Teachers Do*, one of the books that has most influenced my teaching and that of many other educators. One of the main lessons I learned from Ken that day is to start the design of any learning experience (course, class session, workshop, etc.) asking yourself the question: "What should my students be able to do intellectually, physically, or emotionally as a result of their learning?" This led me to discover "backward design," a widely used framework for designing courses and content units proposed by Grant Wiggins and Jay McTighe in their book *Understanding by Design*.[7]

A full treatment of the backward design framework falls beyond the scope of this book, but it essentially involves three stages:

- Stage 1 – Establish your desired results: what are your enduring understandings and learning goals of the lesson, unit, or course?

- Stage 2 – Evidence and criteria: what criteria will you use to evaluate whether your students have achieved the desired results?

- Stage 3 – Learning plan: What learning activities and instructional strategies will you employ?

Apart from the book, there are a plethora of resources on how to implement this method in your teaching. The companion site contains links to some of them.

Using backward design in your teaching has profound implications for how you conceptualize not only the design of your courses but also their execution. It helps you decide what is really crucial for you, and plan and manage your class time accordingly.

Applying this principle to teaching online

- **Establish your learning goals.** Before every online class, I suggest asking yourself: "What are the two or three things that I would like my students to be able to master by the end of this class?" Answering this question can be tremendously helpful in deciding what material to cut from your class plan (before class) and/or what material to skip (during class) if you are pressed for time.

- **Use your learning goals to design your class.** For each of the activities you conduct in class, ask yourself "how is this activity contributing to helping my students master the two or three things I would like them to master?

2.4   Principle 4 – Use online teaching to its comparative advantage

As indicated in chapter 1, transitioning from a physical classroom to a virtual one represents a loss in many dimensions. One tempting way of dealing with this loss (or at least adapting to it) is to try to reproduce what you were doing in your physical classroom to the online environment. This is what newspapers did when they first transitioned to online by simply creating PDF versions of their print product. But as Bharat Anand described, "that model evolved where now publishers have expanded their audiences by leaning in to digital's unique benefits: making news available anytime and anywhere; creating rich multimedia content; drawing on a wide range of sources in addition to staff journalists; sourcing readers for comments and facilitating conversations among them; and personalizing news feeds."[8]

In the same way, teaching effectively online requires going beyond merely adapting the face-to-face experience in the online classroom, to actually using the online medium to its comparative advantages. For live online classes, these include:

- Students have more ways to contribute in class, which could potentially lead to a wider range of participation in your sessions. See part II for more details.

- You and your students can collaboratively create documents (Google Docs, Google Slides) more easily during class. See chapter 6 for more details.

- You can more easily stimulate interactions between students who might not normally interact with each other.

- Students who cannot leave home for some reason (illness, taking care of someone, etc.) can still attend and participate in your class.

- You can more easily record classes to enable students to watch these recordings (in case they missed class) or re-watch them at their own pace.

- Guest speakers are easier to include in your class because they don't need to physically come to your classroom; they can simply join online for part of your class.

As we recognize the comparative advantages of the online environment, it is also important to recognize some of its disadvantages. Some of them – related to the human and emotional consequences of you and your students not being in the same physical space – were acknowledged earlier in this book. I would like to add four more, informed partly by conversations with instructors and students:

- It is easier for students to get distracted. They are just a click away from checking their email, Facebook, Twitter, etc., and you have no way of knowing that this is what they are doing.

- It is harder for you to "read the room" and pick up non-verbal cues, and communication can sometimes get lost (in both directions).

- For you and for many students, learning online is a relatively new experience. There are no long-established norms like they exist in a physical classroom.

- There can be more for teachers to keep track of when teaching a live online class (and all of it is in a screen or two), and it takes a while to adjust to having to manage so many things in a different environment than you are used to.

Applying this principle to teaching online

- Do things to leverage the comparative advantages of online classes, including:

  - Leverage the many ways students can engage in – and contribute to – an online class.
  - Consider recording classes for students who cannot attend.
  - Consider having guest speakers to complement and/or enrich what the students are getting in your course.

- Do things to address the disadvantages of online classes, including:

  - Have a more detailed plan for your online sessions. It's harder to "wing it" during class when you are online.
  - Be more explicit online. For example, when you ask a question on which you plan to spend some time or when you give instructions before sending students to work in groups, it is helpful to write the question(s)/instructions so that all students can see them.
  - Establish norms. It is important to establish norms about how you expect students to engage in your online class. For example, how do you expect them to signal that they would like to talk? Do you expect them to use their video? See sample slide below for a simple set of norms for the beginning of a class, and more details in chapter 12.
  - Simplify or get help if you can. There is a lot to manage when teaching an online live class session in Zoom. See chapter 12 for more details.

Figure 2.2. – Example of slide establishing norms at the beginning of a class

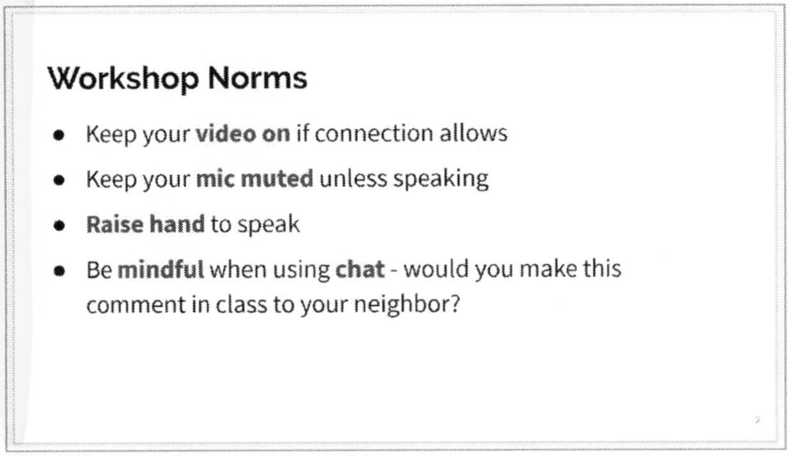

Workshop Norms

- Keep your **video on** if connection allows
- Keep your **mic muted** unless speaking
- **Raise hand** to speak
- Be **mindful** when using **chat** - would you make this comment in class to your neighbor?

Source: "Making the Most of your Online Learning Experience" online workshop at Harvard University organized by Anastacia Kay, Dan Levy, and Teddy Svoronos. April 2020.

2.5   Principle 5 – Teachers are made, not born

If you work hard at your teaching, you know better than anyone that teaching is not just a natural talent. Of course, natural talent helps, but it is the commitment to continuously develop the many skills involved in the craft of teaching that I believe makes the greatest difference. The point might seem trivial, but when you hear expressions like "she is a great teacher" or "what he is doing is not scalable" or "we need to fire the bad teachers," you realize that engrained in the attitudes of many people is the notion that you are either good or bad at teaching and that this is immutable.

Learning to teach online requires deploying skills you already have and developing new ones. Perhaps because of the sense of urgency in which many of us transitioned to online teaching in the spring of 2020, I saw too many colleagues approach it with a sense that everything needed to go perfectly on day one. This approach might provide us with useful aspirations but unrealistic expectations. When Harvard Business School transitioned to remote teaching in the spring of 2020, instructors immediately focused on creating what they called a "Minimum Viable Classroom" (mirroring the term "Minimum Viable Product" often used in the software industry when creating a new application). They worked very hard to create a great first class when they came back from spring break. But they knew it was not going to be their final product. And given their commitment to excellence in teaching and to continuous improvement, they will undoubtedly offer a better learning experience this fall.

This strikes me as a good way to approach your online teaching journey. It won't be perfect on day one. But think of this process in the same way that you think about the process you employed to develop any of the skills you now excel at. Your optimal process will likely differ from mine, so I hesitate to offer advice here, but below are three approaches that I think could be helpful as you develop your online teaching skills.

Applying this principle to teaching online

- Practice a lot

  o It takes a while to get used to the different tools (participants list, chat, sharing screen, etc.).
  o If possible, try different approaches with friends, colleagues, or family members before you try them in your online classes.
  o This will help you relax and focus more on helping students learn.

- Observe others teach

  o It's easier than ever. All you need is an invitation to a colleague's class with a Zoom link!
  o You will learn a lot, particularly if you can debrief with your colleague.
  o You will learn what it feels like to be a student in a live class session. This will help you see the need for interaction. It will also help you get a sense of what seems effective and what does not from the perspective of being a student. This will help you discover approaches you had not even considered.

- Experiment and learn from your experimentation

  o Try new approaches or ideas in your online teaching.
  o Ask a colleague to observe you teach and give you feedback.

---

o   Collect student feedback and evidence about student learning to help you assess how your ideas went in practice.

o   Reflect on what went well and what could be improved.

Chapter Summary

- Think about what pedagogic principles underlie your teaching practices and how best to apply or enact these principles in your live online classes.

- The key pedagogic principles underlying the practices and advice advanced in this book are:

  (1) Be student-centered.
  (2) Plan for active learning.
  (3) Begin with the end in mind.
  (4) Use online teaching to its comparative advantages.
  (5) Teachers are made, not born.

PART II - WAYS YOUR STUDENTS ENGAGE

There are five key channels through which students can communicate using Zoom (or other similar applications) to engage in a live online session: they can speak, vote, write, work in groups, and share work. These channels are not mutually exclusive. For example, students can share work or speak while working in groups. And more importantly, these ways of engagement can often be combined within a segment of the class. For example, after you ask students to vote, you could immediately assign them to work in groups to discuss their vote. Before thinking about how to combine and integrate these channels, I think it's helpful to think explicitly about how students can engage in each channel, as that can be a useful guide for planning your classes. The five chapters in this section address each of these channels.

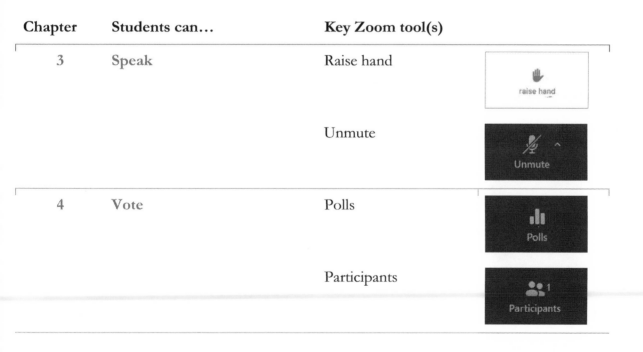

Chapter	Students can...	Key Zoom tool(s)	
3	Speak	Raise hand	
		Unmute	
4	Vote	Polls	
		Participants	

5	Write	Chat	
6	Work in groups	Breakout rooms	
7	Share work	Share screen	

Chapter 3 – Speak

If your class has over a dozen students, you will likely want them to have their microphone on mute at all times except when they are addressing the class. Otherwise, everyone will get distracted as soon as anyone has some noise in their environment (keyboard clicking, dog barking, phone ringing, etc.). For a student to participate in a live online class, the typical sequence is:

- Students raise their hand using Zoom's "Raise Hand" feature (see picture below).

- You select a student who raises their hand and call on them.

- They unmute themselves.

- They state their comment/question.

- They mute themselves.

Figure 3.1 – Student uses two-step process to raise hand in Zoom

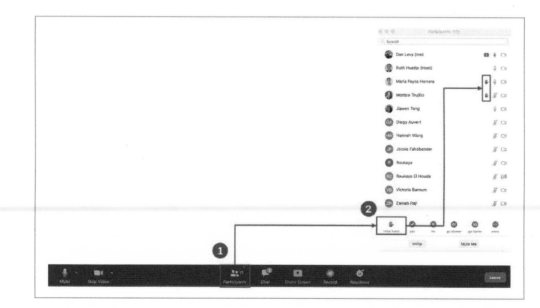

This process (like other online teaching processes) takes a little longer than it would in a physical classroom, and it takes some time for students and instructors to get used to it. Moreover, even without the use of the "Raise Hand" and "Mute/Unmute" features, the mere fact that the conversation is taking place over the internet makes dialogue feel less natural because the internet sends audio and video via chopped-up bits through different channels to the recipient, and then reassembles them. This tends to disturb the "No gap, no overlap" conversational rule that most cultures observe[9].

My colleague Dutch Leonard pointed out that another difference between dialogue in a physical classroom versus a virtual one is that in a physical classroom you can more easily use verbal and non-verbal cues to signal to a student that you want them to bring their comment or question to a close. This means that engaging in dialogue online tends to be more challenging and potentially more time-consuming.

All of this implies that conversations often take longer in a virtual classroom, which has led several of my colleagues to observe that you should cut how much material you plan to engage with in an online session. Some suggest that you aim to get through 80% of the material online that you would get through during the same amount of time in a physical classroom. On the other hand, there are some things (like quickly eliciting student views through the chat function, sending students into breakout groups with different classmates, or bringing the class back to order after working in those breakout groups) that can be done more efficiently online. The bottom line seems to be that you should not assume that you will be able to engage with the same amount of material online as when you teach in person, and that your pedagogic approach and style will likely determine how much less you will be able to engage with virtually.

To decide which parts of your class plan to cut, focus on your learning objectives for the session and think hard about two questions: (1) Is this essential to achieving the learning objectives?, and (2) Will this benefit greatly from us being together in the live session? Sometimes, the material you plan to address in class might be available in another format – like a newspaper, journal article, or YouTube video – that you can post or distribute before or after class. Chapter 10 provides some guidance on how to decide what material should be addressed in the live sessions and what should instead be oriented to learning outside of live classes.

3.1   How do I get my students to participate?

You might be in an environment where students don't tend to participate in class. Or maybe you have some students raising their hand a lot while others rarely do. If you want to have a more inclusive classroom and a richer discussion, here are some things to try:

- Ask good questions. See section below.

- Do not answer your own questions. Wait for students to participate. Silence is one of the most powerful and underused tools at your disposal.

- After you ask a question, pause for at least 10 seconds before you call on someone. This allows every student in the class more time to develop their own thoughts, and usually results in more hands going up, which gives you the opportunity to bring in a wider range of views. This technique is sometimes referred to as "wait time" and has been shown to have many benefits for student engagement and learning.[10]

- After you pose a question to the class, say explicitly: "I would like to see more hands up before I call on someone," or "Let's make sure we have five hands up before I call on anyone."

- After you ask a question, say explicitly: "I would love to hear from someone who hasn't participated."

- Cold call (that is, without telling them in advance, call on someone who did not raise their hand). See "In Practice" below.

- Warm call (that is, with some advance warning, call on someone who did not raise their hand). See "In Practice" below.

- Call on a student who made an interesting comment on the chat.

- Keep track of who has participated and who has not, and be deliberate about whom you decide to call on. You might consider using a tool like <u>Teachly</u>, which is aimed at helping faculty teach more effectively and inclusively (full disclosure: I am one of the founders).

3.2　How do I ask good questions?

Asking good questions is critical to generating good discussions in class and ultimately learning. A full treatment on what constitutes a good question and how to manage class discussions is beyond the scope of this book, but here are some guidelines based on the Center for Teaching and Learning at Washington University in St. Louis:

- Make sure your questions are aligned with your learning goals for that class.

- Aim for direct and specific questions.

- Ask questions throughout class and ask only one question at a time.

- Consider asking open-ended questions.

- When a student answers a question, be intentional about how you respond.

  o Show interest in their answer with verbal and non-verbal cues.
  o Consider asking a follow-up question to clarify, probe, or expand.
  o Invite other students to react (agreeing, disagreeing, or expanding).
  o Be rigorous with the answer but gentle with the student.

Please see the companion site for a few resources on asking good questions and managing discussions in different types of classes.

In Practice 3.1 – The art of cold calling

Mike Toffel teaches Operations at the Harvard Business School, where cold calling is widely used. He notes that cold calls can help ensure that students are prepared for class because they know there's a real chance they will get called on even if they don't raise their hand. But he cautions that there are tradeoffs to cold calling. While you can use it to nudge those students who haven't participated in a while or whom you know have experiences that can add important insight, calling on students who aren't raising their hand risks lowering the quality of conversation – and eroding the motivation of those who *are* raising their hand. That said, Mike uses cold calls throughout the class period, sometimes even cold calling students who participated just a few minutes earlier as a way to keep everyone paying attention and ready to respond as well as to cultivate careful listening skills and a ready-to-contribute mindset.

In Practice 3.2 – The art of warm calling

Many instructors want to have a vigorous discussion in class but dislike cold calling. One effective way of calling on students who haven't participated much is to devote a little time the day before class to thinking about possible connections between the students' backgrounds and the class topic. Below are a couple of illustrative examples.

My Harvard Kennedy School colleague Julie Wilson is a sociologist who has a knack for customizing her teaching plans to the students in her class. When she was recently preparing to teach a case, "Our Piece of the Pie," which takes place in Hartford, Connecticut, she did a quick search of her students' backgrounds in the student roster and found out that one student was from Hartford. She wrote to the student and said she was going to start class by asking her to briefly describe the city so that participants could better understand the context in which the case unfolds. The next day, the student came to class prepared and in a couple of minutes painted a picture of the city that helped ground the ensuing discussion. See box below for the e-mail Julie sent the student.

In my own classes, I often send an online module or survey/quiz for students to complete before class and then leverage their answers to stimulate a discussion during class. A typical sequence in the pre-class survey is a multiple-choice question followed by another question asking students to justify the choice they selected. Sometimes I simply pose a question that I plan to ask in class but that I want students to start thinking about beforehand. In class, I might pose the same question, show a student's answer as a quote (sometimes with the student's name), and invite that student to initiate a conversation or debate. See example below.

Sample e-mail to prepare a student for warm calling

From: "Wilson, Julie Boatright"
Date: Monday, May 18, 2020 at 6:49 AM
To: [Redacted]
Subject: Welcome to the Using Evidence program

Welcome to the Using Evidence program. I just wanted to let you know that I may reach out to you in our discussion of Our Piece of the Pie this morning to say a few words about Hartford.

Those of us who live in New England know it as both the headquarters for many large insurance companies – a site of quite spectacular buildings as you drive through on the interstate – but also a place of considerable poverty and hardship.

While we won't spend much time discussing the background information because we need as much time as possible for laying out a framework for the week, we almost never get to take advantage of someone who actually knows Hartford to help us set the stage!

I hope this works for you. I'll check in before the class.
Julie

Figure 3.2: Bringing into live session a student's response to a pre-class quiz or survey

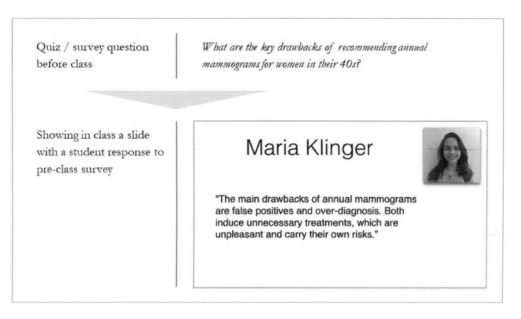

| Quiz / survey question before class | *What are the key drawbacks of recommending annual mammograms for women in their 40s?* |

| Showing in class a slide with a student response to pre-class survey | **Maria Klinger**

"The main drawbacks of annual mammograms are false positives and over-diagnosis. Both induce unnecessary treatments, which are unpleasant and carry their own risks." |

 **Tech Tips**

- In Zoom, students who use the "Raise Hand" feature appear at the top of the participants list (immediately below the hosts and co-hosts) in the order in which they raised their hand.

- Hands remain raised unless someone "lowers" them. Students can lower their hand simply by clicking on same hand-shaped icon a second time. Hosts and co-hosts can also lower the hand of any participant. It can be distracting and inefficient to call on a student who has their virtual hand up unintentionally, so try to set up processes and norms to avoid this. For example, you can ask a member of your teaching team to lower the hands of students who have just made a comment.

- Sometimes a student starts to speak, and you and the rest of the class cannot hear well. This typically happens for reasons related to internet bandwidth. At this point, you need to decide how much time to let pass before you intervene. I typically wait for about 10 seconds and if it hasn't resolved by then, I ask the students to write the comment on the chat, move on, and then come back to the chat to address the comment a few minutes later.

- The phrase "you are on mute" is probably the most widely uttered phrase in Zoom. For students new (and even not so new) to Zoom, it takes a while to get used to the routine that they have to unmute themselves before saying something to the class. If you call on a student and they don't respond within a few seconds, you might want to intervene and say "you might be on mute" so as not to waste precious class time on this. If you want to find out whether a particular student is on mute or not, you can look at the microphone-shaped icon next to their name on the participants list (right in between the hand-raising icon and the video feed icon).

- While I understand the need for students to be on mute during class, the default "everyone on mute" seems very unnatural and constitutes one of the hardest things to adapt to when moving from a physical classroom to a virtual one. Humor does not travel as well. More generally, you and your students cannot "hear the classroom" and miss some of the things that sometimes make a class in a physical classroom such a human experience, as described in chapter 1. You miss the laughter, the quick reaction on one side of the room to a comment from a student on the other side, the occasional chuckle, etc. I don't have a solution to this conundrum but hope that we can evolve into a different default soon with the help of technology or some creative ideas.

Sample checklist for calling on students

<u>Before class</u>

❏ If you are planning to teach online a class you have taught in person previously, plan your time carefully. Consider cutting some of the material (or at least having a plan for what you would skip if pressed for time).

❏ Plan what questions you want to ask in class to achieve your learning goals.

❏ Have a plan regarding which students you want to call on during class.

❏ Consider reaching out to specific students ahead of time, so you can warm call them during class.

<u>During class</u>

❏ Always have Zoom's participants list open and monitor it so you can see if/which students are raising their hand.

❏ Resist the temptation to call on the first hand raised. Be deliberate about whom you call on.

❏ Put the list of students you want to call on in class somewhere visible.

<u>After class</u>

❏ Reflect on the kinds of conversations you were able to stimulate in class and the diversity of students and viewpoints reflected in these conversations.

❏ Write lessons learned for your next class!

Chapter Summary

- Conversations tend to take longer in a virtual classroom than in a physical one. Take this into account when you plan your class.

- Set norms for students to raise their hand during class.

- There are many techniques to ask good questions and help stimulate students to participate in class. Use them!

Chapter 4 – Vote

If you have a class with more than a dozen students, one very effective way of aggregating your participants' views and helping you understand where the class is with respect to a particular topic or question is to poll them. Polling has many benefits,[11] including:

- Nudging students to actively engage with the material

- Encouraging participation from all students

- Providing you with a more objective view of students' perspectives on a topic

- Helping both you and your students check for understanding during class

- Helping you adapt your teaching plan to where your students are in the class

For a wonderful and thorough description of the use of polling in teaching, I highly recommend Derek Bruff's book *Teaching with Classroom Response Systems* (2009). While the book was written with a physical classroom in mind, its key principles and wonderful examples make it an excellent guide for the use of polling in online teaching. For a shorter guide on the use of polling devices, take a look at a post he wrote for the Center for Teaching he directs at Vanderbilt University. Links to these resources are available on the companion site. The table below gives you a sense of the kinds of questions you can ask when polling your students.

Table 4.1 – Types of questions in polling

Type of question	Example
Recall	What are the basic units of life called? A. Bacteria B. DNA C. Cells D. Genes

Conceptual understanding	You roll a die twice. What is the probability of getting at least one "6"?
	
	  A. 1/6
	  B. 2/6
	  C. 1/36
	  D. 11/36
	  E. None of the above
	  F. I don't know

Application	Evaluators are studying the impact of a microfinance program using a randomized controlled trial. Suppose there was a drought in the region where the study was being conducted.
	How does this drought affect the credibility of the study?
	
	  A. Would not affect credibility
	  B. Would somewhat affect credibility
	  C. Would substantially affect credibility
	  D. I don't know

Student perspectives	If you were in the shoes of the protagonist of the case, would you merge with the other company?
	
	  A. Yes
	  B. No

Student preferences	Which of the following topics would you like to discuss in our midterm review today?
	
	  A. Utility functions
	  B. Public Goods
	  C. Externalities
	  D. Taxes
	  E. Subsidies

Confidence level	How confident are you in your answer to the previous question?
	
	  A. Very confident
	  B. Confident
	  C. Somewhat confident
	  D. Not confident

| Feedback/monitoring questions | How difficult was the assignment you just turned in?
  A. Very easy
  B. Easy
  C. Appropriate
  D. Difficult
  E. Very difficult |

Sources: Typology adapted from Bruff (see <u>post</u>). Correct answers for first two questions are C and D.

In Practice 4.1 – Use of polls

In 2008, I participated in a teaching workshop on the use of polling devices and peer instruction led by Harvard physics professor Eric Mazur. This workshop and several subsequent conversations with Eric have profoundly influenced the way I teach. Then, in the fall of 2010, I had an aha moment. While teaching a session on probability, I used a polling device to pose what I believed to be a warm-up question, expecting at least 80% of my students to arrive at the right answer. But when the poll results came in, I was stunned and stood in silence in front of the class for over a minute. Only 17% of participants had answered the question correctly. This was a humbling experience for me and made me realize how clueless I can be about where my students are. Since that day, I have relied on polls frequently in teaching to assess student understanding, take the temperature in the room, gauge levels of interest on different topics or activities, ask sensitive questions, and many other purposes.

The rest of this chapter is organized as follows. First, I discuss whether polls should be anonymous, a key decision to make when designing polls. For both anonymous and non-anonymous polls, I discuss how to implement them in Zoom, their key relative advantages, what to do after polling, and some examples of real instructors using them. I then explore how to use polls in small classes. Finally, I wrap up with some tech tips, a checklist for using polls, and a chapter summary.

4.1   Should polls be anonymous?

A key decision to make is whether you want the students' votes to be anonymous. An anonymous poll allows you to see how the students voted in aggregate (e.g., 23% of students voted for option A, 54% voted for option B, etc.) without identifying which student voted for which option. A non-anonymous poll allows you to see both the aggregate results and how each student voted individually. The method for non-anonymous polls suggested in the table below also allows students to see other students' votes.

Table 4.2 – Key features of anonymous vs. non-anonymous polls

Type of polling	Suggested Zoom tool	Key features
Anonymous	Polling (in Zoom's main bar)	Can see aggregate results but not individual votes.
		Can decide when to share poll results with the students.
		Should create poll in Zoom before class.
Non-anonymous	Buttons at bottom of participants list (Yes, No, Go slower, Go faster, etc.)	Can see both aggregate results and individual votes.
		Students can also see how other students vote.
		Can create poll before class but also on the fly.

Figure 4.1 – Reporting of results in anonymous vs. non-anonymous polls

Anonymous polls

Non-anonymous polls

You can use a mix of anonymous and non-anonymous polls in your teaching. Below is a description of the advantages and disadvantages of each to help you decide when to use each kind of poll.

<u>Anonymous polls</u>

In Zoom, you can use the "Polling" feature for anonymous polls. Below is a table explaining the different tasks associated with this feature, and information on when to do them and who can do them. Note that the poll results are visible to you when you end the poll, and they only become visible to the students if you decide to share the results. This means you can check the aggregate poll results in private and decide if and when to share the results with your students.

Table 4.3 – Tasks in Zoom's "Polling" feature for anonymous polls

Task	When to do it	Who can do it
Create poll	Before class	Person who creates the Zoom session
Launch poll	During class	Host or co-host
End poll	During class	Host or co-host
Share poll	During class	Host or co-host
Re-launch poll	During class	Host or co-host

For detailed tutorials on how to perform each of these tasks, visit the <u>companion site</u>.

To launch an anonymous poll during your live class, you click on the "Polling" icon on Zoom's main toolbar, which will then give you the option of launching any of the poll questions you created for that session. Select the poll question you want to launch and then click "Launch Poll." See picture below. If the "Polling" icon does not appear on your Zoom bar, you need to add it (please see "Tech Tips" section below for instructions).

Figure 4.2 – Launching anonymous poll from Zoom's main toolbar

An anonymous poll has the advantage that students are more likely to express their real views about the question you are asking them. This is particularly important with questions that are sensitive or where students might feel nervous about expressing unpopular views. But it is also important for other questions where students who are unsure of their answer might deliberately or inadvertently side with the emerging majority (or with specific students who are known to be strong on the subject), which might lead you to inaccurately assess your students' level of understanding[12]. Finally, by making a poll anonymous, students don't know how others voted until you share the results. My general advice is that if you want to have a vigorous debate, you should not share the poll results until after the debate is over, so students are not unduly influenced by how others voted. The possible exception to this advice is if the vote is very split (i.e., top 2-3 choices gather a similar percentage of votes), in which case showing the results might spur people who were unsure of their answer to participate.

What should I do after I launch a poll in class?

After you launch an anonymous poll and see the results privately, you can take a number of actions. Below is a table summarizing some of your options. Some of these come from Mazur (1997) and Bruff (2009), while others are adaptations to my own teaching practices. You can of course mix some of these options, come up with variants, or use entirely different ones; this list is just meant to stimulate your thinking. If you are interested in seeing some of these actions in practice, check the companion site.

Table 4.4 – Some possible courses of action after you launch a poll and see the results

	Typical Use	Course of Action			
A	Initial poll results show class is very split and students might benefit from small group discussion before wider debate	Small group discussion	Re-launch poll	Class-wide discussion	Share poll results with students
B	Initial poll results show class is very split and you want to have a debate	Share poll results with students	Class-wide discussion		
C	Initial poll results are not split (i.e. majority of students voted for one answer)	Share poll results with students	You explain the answer or invite one student to argue for the majority answer		
D	Use poll results just to know where students are and help you steer the discussion	Class-wide discussion			
E	You want to see how much students change their view as a result of the class-wide discussion	Class-wide discussion	Re-launch poll	Class-wide discussion	Share poll results with students

If you are new to polling in your teaching, one small step to get you started is to prepare a poll for your next class, launch it, see the results, and invite the class to discuss/defend their answer. If you are having difficulty getting students to participate, you might nudge some of them by saying something like, "40% of you voted no. Can one of you share with the class why you think that?"

One disadvantage of using the polling feature of Zoom is that to create a poll while in class you have to go to your Zoom account on the web, type the question, type the possible answers, return to Zoom, and launch the poll. This can take time. For this reason, it is better to create the poll before class. This has to be done by the person who created the Zoom meeting. One way to address this disadvantage of anonymous polls is to include in your list of polls for every class a generic question that simply has as possible answers A, B, C, D, etc. (see image below). This way, if you come up with a question during class that you want to poll your students on, you can simply launch the generic poll.

Figure 4.3 – Generic poll

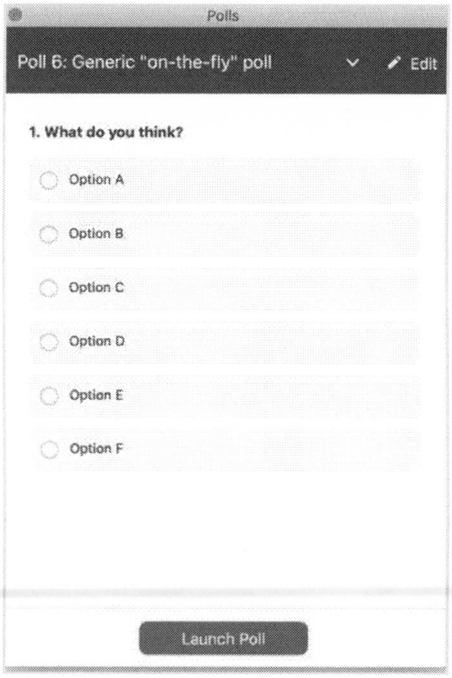

<u>Non-anonymous polls</u>

Using a non-anonymous poll has the advantage that by knowing how each student voted, you can follow up the vote by asking specific students why they made the choice they made. This could help generate a rich debate and make it easier for you to call on students who haven't participated much. It also has the advantage that it tends to be a little quicker and you can nudge the students who haven't voted to vote (e.g., "Ricky, I noticed you haven't voted. Could you please do so?").

The most common way to implement a non-anonymous poll in Zoom is to ask a question and have students vote by pressing one of the buttons at the bottom of their participant list. Once they do so, their vote will be reflected in the participant list so everyone can see how everyone else voted.

The typical use case is for yes/no questions. But if you want to use a non-anonymous poll with more than two possible answers, you could ask the students to use other buttons at the bottom of the participants list. For example, if you think the answer is A press "yes", B press "no", C press "go slower," and D press "go faster". See picture below.

Figure 4.4 – Non-anonymous poll with more than two possible answers

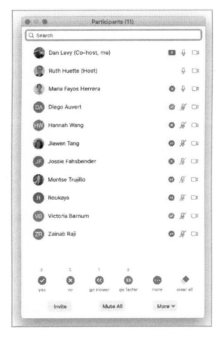

You can create non-anonymous polls in advance or on the fly. Regardless of when you create them, it is helpful to be very explicit with students about how to vote. For yes/no questions, it is fairly obvious for students to know what to do. But even then, keep in mind that some students might not have heard the question well, so consider having a visual cue (e.g., a slide) reflecting what the question was. Furthermore, for questions where the mapping between the possible answers and Zoom's buttons is not obvious, being explicit is crucial. In my own teaching practice, I tend to show participants a slide that indicates the mapping between the possible answers and Zoom's buttons. See example slide below.

Figure 4.5 – Sample slide for non-anonymous polls

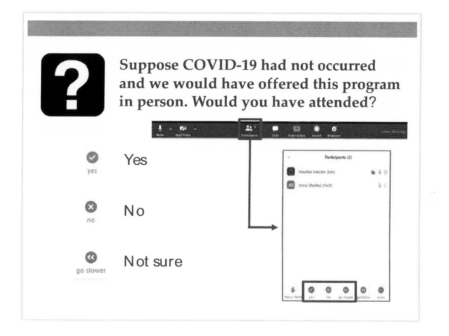

In Practice 4.2 – Using non-anonymous polls

Rebecca Nesson, Associate Dean of the Harvard College Curriculum, teaches computer science and math courses at Harvard and uses non-anonymous polls in interesting and flexible ways. She frequently asks a question during class followed by "If you think the answer is X, press Yes; if you think the answer is something else, press No." After students have voted, she deliberately picks a student from the list and calls on him or her to justify the choice (e.g., "Jamie, I noticed you voted yes. Could you please explain your reasoning to the class?"). She thinks this works better for her than traditional cold calling (where you just call on a student to answer a question immediately after you ask the question) because she knows that the student has already thought about the question and will likely have something to say.

Rem Koning teaches strategy courses at the Harvard Business School and uses non-anonymous polls to generate vigorous debate in his case discussions. In many business school cases, the protagonist has to make a decision (e.g., to buy or not to buy, to merge or not to merge, etc.). In a recent online class, he asked the students whether the company featured in the case had to do "more of the same" (press "Yes") or "do different" (press "No"). He then wrote on a flipchart behind him, where he titled two columns "More of the same" and "Do different" and proceeded to call on specific students to submit arguments to support each of the choices (e.g., "Veronica, you voted for doing different. Can you explain why?") and then wrote the students' comments in each of the two columns. I found Rem's adaptation to the online environment of what traditionally happens in a business school classroom to be creative and resourceful. See picture below.

Figure 4.6 – Harvard Business School professor Rem Koning adapting traditional polling methods

4.2   Using polls in small classes

If you teach a relatively small class (fewer than twenty students), you can poll students in ways other than the ones described above. For example, you might pose a multiple-choice question, and ask students to show one finger in their hand if they think the answer is 1, two fingers if it's 2, and so on. Or you might have students show you cards of different colors (red for the first answer, blue for the second one, etc.). You can then quickly explore the video feeds of your students in Gallery View to give you a sense of the aggregate vote and to know how specific students voted. The advantage of doing it this way over using Zoom's native polling tool is that you can probably do it a little more quickly.

In Practice 4.3 – Using polls in small classes

Shoshanna Kostant is a math teacher at Brookline High School in Massachusetts. She splits her class into two groups to be able to conduct smaller sessions with each of them, and combines polling with other teaching approaches in a very creative manner.

When she wants to do a quick non-anonymous poll, she asks her students to signal their response using their hands (e.g., if you think this series converges, show your hand in form of a "C"; if you think it diverges, show your hand in form of a "D"). This allows her to see how her students voted very easily and quickly by just looking at the video feeds of her students in Gallery View.

When she wants to do an anonymous poll, she asks her students to chat their answer just to her, privately. This has the advantage that she does not need to give her students a list of possible answers that she would have to think of beforehand, and allows her to identify possible misconceptions that she might not have even been aware of. She can simply say, "Solve this problem and send me your answer through the private chat" and then examine her students' answers and call on any of them to explain their response.

Tech Tips

- If the "Polling" feature does not appear on your Zoom's main toolbar, you need to add it. Please note that this feature is currently not available on Zoom's basic (free) plan. To add the "Polling" feature to your Zoom's main toolbar, follow these steps:

  o Login to your Zoom account on your browser. The web address will depend on your institution's setting.
  o On the left navigation bar, click on Settings. See image below.

o   Click on sub-menu "In Meeting (Basic)" and scroll down until you find the item related to polls. Make sure it is turned on.

- To launch polls using Zoom's built-in poll feature, it is best to create the polls in Zoom **before** your class session. Once in class, you can launch any of the polls that are pre-loaded for that class.

- Only the person who sets up the Zoom meeting can create polls for that meeting/class. So if someone else is setting up your Zoom class sessions or you are invited to give a guest class, you should coordinate with the person setting up the Zoom meeting(s) so they create your poll(s) in Zoom in advance of your class. Once the class begins, both hosts and cohosts can launch polls, share results, etc.

- When you share the poll results, your students see a window where the results appear in front of their Zoom screen. Once you stop sharing the poll, this window will disappear.

- Zoom's native poll capabilities are fairly basic. You can only create two kinds of questions: multiple choice (respondent can select only one answer) or multiple answers (respondent can select one or more answers). There are polling applications (such as polleverywhere, mentimeter, etc.) that offer more capabilities (such as questions where respondents can type words and a word cloud gets formed). You could use these polling applications alongside Zoom, and ask your students to visit an external site to answer the poll, perhaps by sending them the link in the chat when you want them to do so. The disadvantage of using another application is that students are already managing multiple windows in a Zoom call (gallery view, participant list, chat, any program they use to take notes, etc.), so using a poll application outside of Zoom adds one more window to manage. If students are managing everything from a laptop computer and/or they are not very technologically savvy, this disadvantage can be particularly salient. In sum, Zoom's native poll capabilities are more convenient but less powerful. Although I was a big user of a more sophisticated polling solution in the physical classroom, I have tended to rely on Zoom for online polling. My hope is that in the near future Zoom's polling capabilities will get better, and/or the more sophisticated polling applications will build integrations with Zoom that would allow students to vote without leaving the Zoom interface.

- You can also use Zoom's polling feature to conduct a non-anonymous poll. In fact, this is Zoom's default. But what Zoom calls a non-anonymous poll is one in which you can download a report **after** class that would show you how each student voted along with their names (assuming that the students registered for the Zoom meeting). This report is not available during class, so it might be helpful to analyze what happened after class but not to make decisions in class such as whom to call on. For this reason, I recommend that if you want to conduct a non-anonymous poll, you ask students to vote by using the buttons at the bottom of their participant list (as described earlier in this chapter).

- If there are not too many students in your class, you can use non-anonymous polls to assign participants into breakout rooms. For example, you could say, if you want to do activity A press "yes", B press "no", C press "go slower," and D press "go faster". For details, please see chapter 6.

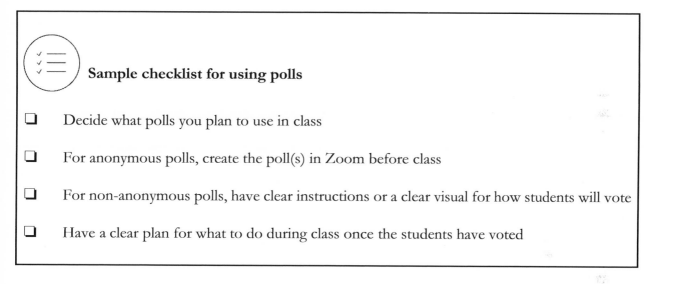

Sample checklist for using polls

❏   Decide what polls you plan to use in class

❏   For anonymous polls, create the poll(s) in Zoom before class

❏   For non-anonymous polls, have clear instructions or a clear visual for how students will vote

❏   Have a clear plan for what to do during class once the students have voted

Chapter Summary

- Polls can be a very helpful tool to engage students, assess where they are, and teach in a more flexible manner.

- For each poll you plan to use, consider whether you want it to be anonymous or not. There are tradeoffs in this choice and my recommended approach to implement these in Zoom differs (Zoom's native polling tool for anonymous polls, and Zoom's buttons at the bottom of participant list for non-anonymous polls).

- Be deliberate about what you plan to do once you learn the results of the poll. Key options include leading a class-wide discussion, asking specific students to justify their vote, and assigning students to work on the question further in breakout rooms. Consider the possibility of repolling after any of these activities to see if the views of the class changed.

Chapter 5 – Write

While voting is very useful to quickly aggregate student views, you sometimes want to know about what is on your students' minds in a less structured way than using a multiple-choice poll. Zoom's chat feature enables students to write what they are thinking as the class progresses and/or in response to a prompt from you. If you have ever dreamed of teaching in a classroom where you could see in thought bubbles what is going on in your students' minds, you might want to consider the use of chat in your online live classes.

Figure 5.1 – Thought bubbles

Image Source: Classroom Doodle Diva.

That said, the use of chat in a live online class is controversial. Some instructors encourage their students to use it freely, many place some restrictions (like whether everyone or just the instructor can see what students are writing), and others simply ban it. Students also seem split, with some finding it a useful way to participate, and others finding it too distracting. The table below points to some of the main advantages and disadvantages of using chat in a live class session. The last row assumes that you are using Zoom's most liberal (and currently default) setting, which allows students to see what other students are writing.

Table 5.1 – Key advantages and disadvantages of using chat during class

Key advantages	Key disadvantages
+ Allows you to quickly see what's on your students' minds more efficiently than in a physical classroom	− Can be distracting for you
+ Helps you uncover issues that you need to address / clarify during class	− Can be hard for you to monitor while you are trying to manage many other things during class
+ Can help resolve technical issues quickly	
+ Provides an additional venue to contribute to class for students who might be hesitant to speak in class	
+ Allows more students to actively engage and participate in class, just like with voting	

If students can see what other students are writing:

+ Helps students learn from each other without intervention from you	− Can be distracting for students
+ Students can help answer other students' questions	
+ Can bring a richer set of perspectives into the class	
+ Creates community by allowing students to interact with and learn from each other	

The table above suggests that there are more advantages than disadvantages to using chat, but this can be deceiving, because the disadvantages are big ones. I address them below, but before doing so, let's examine ways you can use chat in your live online class.

5.1   Main uses of chat

<u>Use chat to have students respond to your prompt</u>

Asking your students to write in the chat can be a very efficient way to quickly find out what they think about a specific topic or issue. In a traditional classroom, you might ask a question and then get a few students to answer verbally. Because this process is sequential (i.e., you don't want all students to speak at the same time), it would likely take more time and involve fewer students than using the chat tool (where multiple students can write at the same time).

Using chat allows you to quickly look for interesting patterns in the students' responses, pull out a particularly interesting comment to discuss verbally, and ask a student that you want to nudge into participating in class to expand on their chat comment. In a traditional physical classroom, this would all be much more difficult and time-consuming.

Below are some illustrative examples of the kinds of questions for which the chat function would work well. There are of course many more. The key is not to try to find a question that fits the chat function, but rather to substantively think of questions where you would like to know what's on your students' minds and ask yourself whether the chat is the right vehicle for it. In the process of doing so, some questions that you might not have thought about asking in a physical classroom might emerge.

You may want to keep two factors in mind when deciding whether a particular question is a good candidate for chat and, if so, how to communicate the question to your students: (1) how long you expect the typical answer to be (a few words or one to two sentences is ideal in terms of your ability to collect and process the comments quickly); and (2) how many students you expect to contribute (the greater the number, the more structure you need in the question and the shorter the student responses should be).

Table 5.2 – Possible prompts for students to respond using chat

Example of prompt	Goal
When you hear the term X, what are the **first words that come to mind**?	Assess students' initial understanding of a concept/idea, and identify potential misconceptions.
What **examples** come to mind to illustrate Y?	Quick, low-barrier-to-participate question aimed at helping students come up with a broad range of examples that you can leverage in the ensuing discussion.
What **questions** do you have about the material so far?	Help you uncover things that need clarification. Can also help you be more selective about which questions you answer in class.
How would you **apply** this concept/idea to **your own life**?	Help students apply what they have learned. Typically, you would need to give students a little bit more time for this kind of question.
What are the **advantages/disadvantages** of X? What would be the reasons to do X?	Quickly come up with a list that you can use to jumpstart a conversation
What are your **key conclusions** (i.e., takeaways) from today's class?	Help your students do retrieval (a process that leads to learning) and allows you to understand what your students took away from the class (and compare it with what you planned for them to take away).
What did you do **last weekend**?	Foster a sense of community before class.

Consistent with the general theme about the importance of being explicit when communicating with students in live online classes, it sometimes can be helpful to write the prompt in a visual (e.g., slide, whiteboard, etc.) that students can refer to while you ask them to write in the chat. The prompt can also indicate how long you expect their answers to be. See example below.

Figure 5.2 – Slide with question asking students to write in the chat

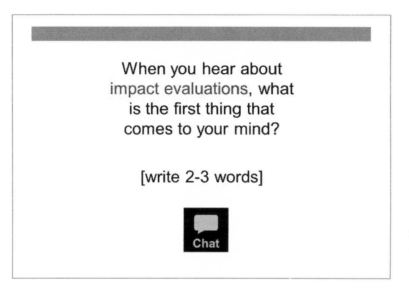

⊙ **In Practice 5.1 – Use of chat in response to your prompt**

Teddy Svoronos, a Lecturer at the Harvard Kennedy School and the person most skilled at using technology to teach effectively that I have ever seen, frequently uses chat to find out what is on his students' minds. In a recent class, he asked his students to write in the chat the methodological and practical criteria relevant to select a particular statistical method. Within a few seconds, he had more than a dozen comments to choose from. He selected some of them, mentioned them verbally to the class, and wrote them on the slide he was showing the students (see picture below). At the end of the class, he asked the students to write their key takeaways from the session, which gave him a better understanding of what the students had learned as well as a starting point for his next class.

Figure 5.3 – Annotating a slide based on comments from the chat (Teddy Svoronos)

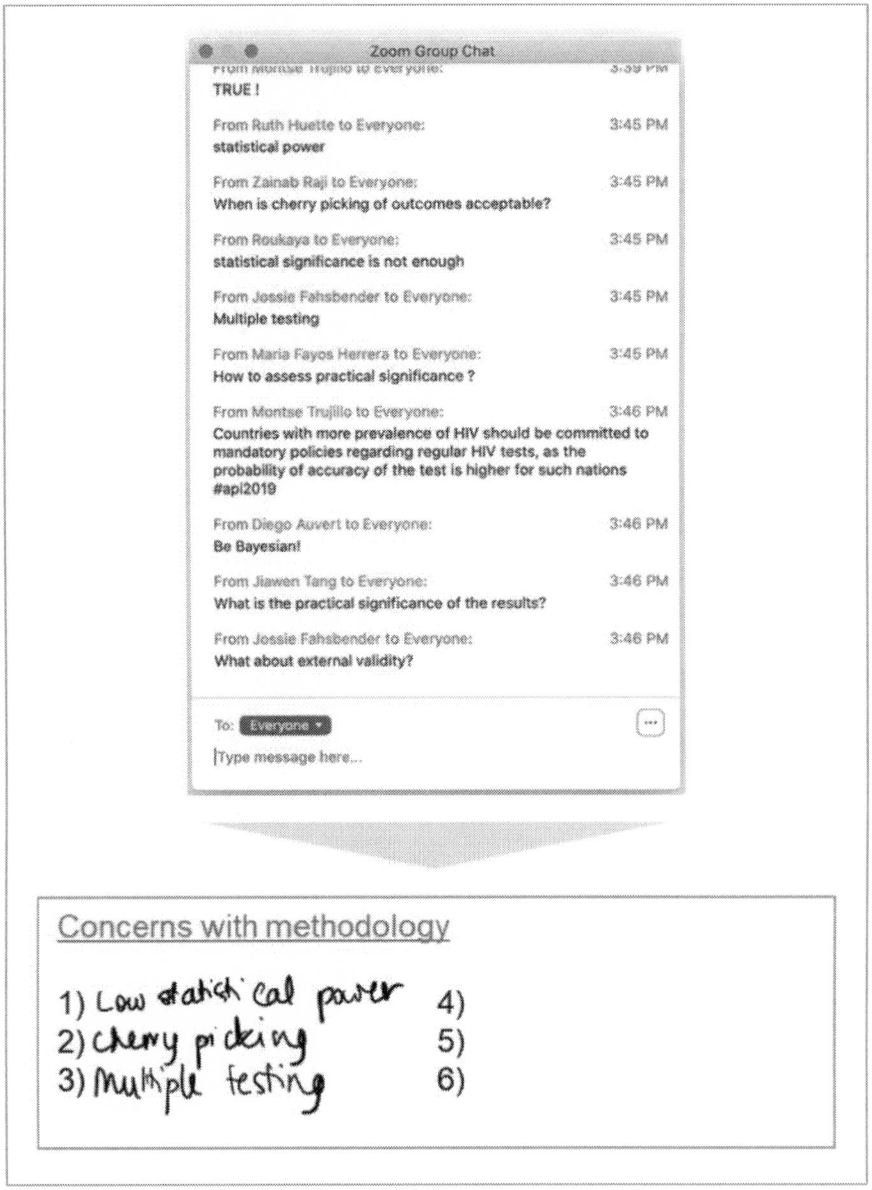

<u>Use of chat to communicate administrative or logistical matters during class</u>

Sometimes chat can be the most efficient and discreet way to communicate things that are relevant to the logistics of running the class. As the instructor, you can use it to communicate messages such as:

- Class will begin in 2 minutes.

- When we begin class, group #3 will be asked to present their key takeaways from last class. Please make sure that you have designated a spokesperson for your group and that they are ready.

- Here is the link to a site that I would like you to look at / work on (e.g., Google Slides for upcoming breakout room).

Figure 5.4 – Rebecca Nesson sends a quick chat to get students ready to start her class

As for students, they might communicate messages such as:

- We cannot see your slides.

- We cannot hear you well.

- Can you repeat the instructions again?

- How much time do we have for this activity?

Notice that some students' questions can be answered by their peers, allowing you to continue without having to pause the class to answer them.

Use of chat in private mode

A few settings in Zoom go beyond the public use of chat described so far, where each comment is visible to all students and the instructor. By default, a student can also write privately to one individual, including to you or one of their peers. You can modify Zoom's default settings so that some of these kinds of messages are allowed and others are not. See "Tech Tips" section below on how to do this.

If you decide to use chat under a certain set of norms (see below), my suggestion is to specify the use of these settings when developing these norms. For example, if you have a person who can help students address technical support issues, you might want to encourage students to chat privately to that person to avoid distracting other students in the class with technical issues that are only relevant to certain individuals.

Use of chat in free-flowing form

Sometimes students use chat to simply express what is on their mind while the class is taking place. Some of the comments are meant to express viewpoints that agree, contradict, or complement points being made verbally or via chat. Some chat comments are then followed by other chat comments that agree, debate, or complement them. When this happens, it can feel like there are two parallel conversations taking place in the virtual classroom: one by voice and a second one by chat. Sometimes students post links in the chat that relate (or not) to one of the two conversations. In fact, there often are *more* than two conversations taking place because sometimes several students write in chat simultaneously, so several conversation threads occur within the chat. These scenarios have made some educators and students oppose the use of chat in live online classes.

Figure 5.5 – Use of chat in free-flowing form leading to several conversations in your classroom

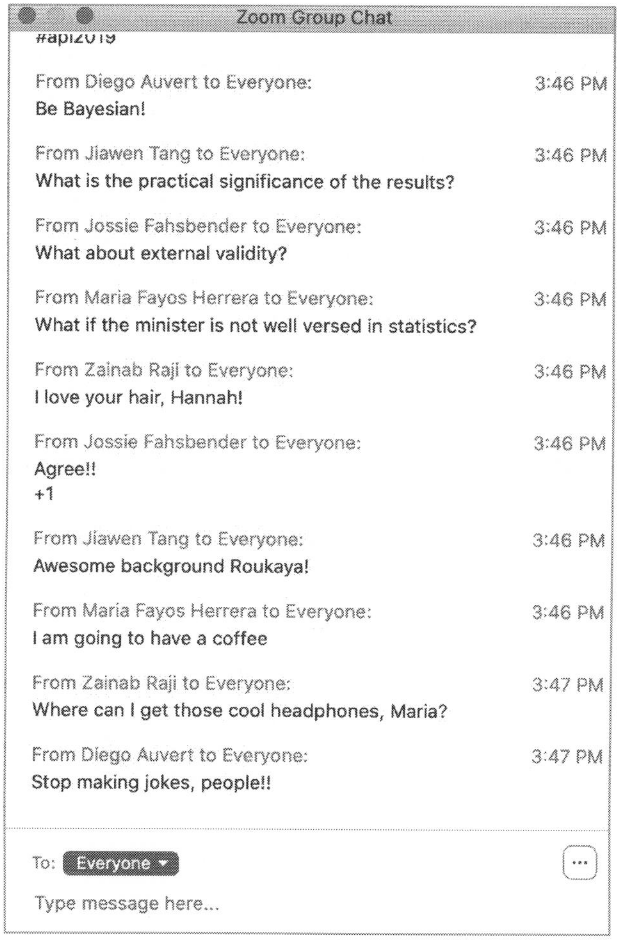

```
●  ●  ●              Zoom Group Chat
#api2019

From Diego Auvert to Everyone:              3:46 PM
Be Bayesian!

From Jiawen Tang to Everyone:               3:46 PM
What is the practical significance of the results?

From Jossie Fahsbender to Everyone:         3:46 PM
What about external validity?

From Maria Fayos Herrera to Everyone:        3:46 PM
What if the minister is not well versed in statistics?

From Zainab Raji to Everyone:               3:46 PM
I love your hair, Hannah!

From Jossie Fahsbender to Everyone:          3:46 PM
Agree!!
+1

From Jiawen Tang to Everyone:               3:46 PM
Awesome background Roukaya!

From Maria Fayos Herrera to Everyone:        3:46 PM
I am going to have a coffee

From Zainab Raji to Everyone:               3:47 PM
Where can I get those cool headphones, Maria?

From Diego Auvert to Everyone:              3:47 PM
Stop making jokes, people!!

────────────────────────────────────────────
To:  Everyone ▾                              (...)

Type message here...
```

5.2   The case against chat

The main case against the use of chat in online classes is that it is too distracting for both the instructor and the students. For the instructor, monitoring the chat while you are teaching can be challenging. In a typical online class, you are paying attention to many things beyond executing your class plan: the video feed of students so you can connect and read non-verbal cues, the list of participants so you can keep track of who is raising their hand, the slides/visuals you are presenting, the clock to keep track of time, and so on. Adding the monitoring of the chat window to this list can seem like a daunting task, especially if you are new to teaching live sessions online and/or if you are not very comfortable with the technology.

One way of addressing this problem is to designate someone else to monitor the chat. The ideal person is a teaching partner or teaching assistant who is familiar with the class content and your class plan, so they can exercise judgment as to what to bring to your attention. If this is not possible but you still would like someone to monitor the chat, you could assign someone (perhaps a student) to play this role and give them instructions on the kinds of issues they should bring to your attention and when. Finally, you could designate times in which you plan to check the chat, and tell the students that you will only be checking it at those times.

 In Practice 5.2 – Having someone monitor the chat

Julie Wilson, a Senior Lecturer at the Harvard Kennedy School, was teaching recently in a program for mid-career professionals. She gave me a list of students she wanted to call on during class and a sense of what she wanted to achieve in that session. At a few points during the class, I interjected with comments such as, "Julie: In the chat, several students are expressing confusion about the term theory of change in this context," or "Julie: Ahmad argued that the theory of change is incomplete." I tried to make sure that the interjections were enriching the conversation and not disrupting her flow, so I interjected most often in response to her pausing and asking, "Dan: Anything in the chat that you would recommend that we address?" This process allowed Julie to leverage the chat without having to monitor it herself.

The second – and, in my mind, bigger – problem with chat is that it can be distracting for students. There is ample research suggesting that most people are not good at multitasking,[13] even though many people believe they are,[14] further compounding the problem. My own experience observing the use of chat is that the conversations in chat can frequently veer off topic. And even if they are on topic, it is cognitively taxing to try to pay attention to two separate conversations. Finally, any links posted on the chat are an invitation to yet another potential source of distraction, as students are tempted to click on those links immediately to view that material.

5.3   The verdict

The decision of whether or not to use chat, and if so how to use it, is a personal one. What I can recommend is that you be deliberate in making this decision and clearly communicate your choice and the reasons behind it to your students. I think your options are as follows:

- Ban the use of chat completely.

- Permit the use of chat under a certain set of norms (when to use it, private vs. public, etc.), and decide how to enforce these norms (e.g., by urging students to embrace them and/or by imposing restrictions by changing Zoom's settings).

- Permit the use of chat freely.

If you are inclined to ban the use of chat completely, keep in mind that students might decide to have their own way of chatting outside of Zoom, in which case they will still be chatting, but you just won't know what they are chatting about! I have seen students do this using WhatsApp groups, Slack channels, etc.

My own view at the moment, which might change over time, is that chat can be very beneficial if you can develop a set of norms that work for you. Below is a sample set of norms you might consider adapting. I also suspect that there will be technological advancements in the next few years that might allow us to address some of the challenges of using chat raised in this chapter.

Table 5.3 – Sample norms for using chat

Stringent set of norms	Relaxed set of norms
• Use it privately for technical support.	• Use it privately for technical support.
• Use it publicly only when I ask you to do so.	• Use it publicly for questions or comments related to the class.

Finally, whatever set of norms you decide to use, there are bound to be some students who will still find it too distracting or dislike using it, but are worried that they might miss something important in the chat. One possible strategy to address their concern is to give these students the option of closing their chat window, tell them that you will verbally announce to the whole class any comments in the chat that are time sensitive and relevant to the whole class, and indicate that you will make the chat transcript available after class (e.g., by posting it in your learning management system). You can copy the content of the chat at the end of class or access it on your Zoom account (see <u>companion site</u> for instructions on how to do this).

Sample checklist for using chat

☐   Establish and communicate norms. Consider changing Zoom's chat settings to help you enforce these norms (see "Tech Tips" section below).

☐   Take action if the norms are violated.

☐   Decide who will monitor the chat. If it's someone other than you, coordinate how and when you want this person to bring chat comments to your attention.

☐   For every question you plan to ask in class, assess whether chat can be a productive vehicle for gathering your students' views.

Tech Tips

- To change Zoom's chat settings once the session has started, simply go to the bottom of the chat window, click on the square with the three periods, and select the option that you prefer. See picture below for how to change settings and table below for Zoom's current settings. Note that you can also change Zoom's chat default settings for all your Zoom meetings by logging into your Zoom account on a web browser, and going into the Settings tab.

Figure 5.6 – Changing the chat settings in Zoom

Table 5.4 – Zoom's current settings

How stringent the setting is	Participants can chat with	Means
Very stringent	No one	You are banning the use of chat. You can still use the chat feature to send messages to your students, but your students cannot send or reply to messages.
Stringent	Host only	Your students can send private messages to you (and to each of your co-hosts) but the messages will not be visible to anyone else (in particular other students, unless you assign them as co-hosts).

| Liberal |  | Everyone (publicly) | Your students can write messages in the chat that will be visible to you and everyone else in the session, but they cannot send private messages to each other. |

| Very liberal | | Everyone (publicly and privately) | Your students can send messages that are visible to everyone else and also private messages to other students. |

Chapter Summary

- Chat is a very quick and efficient way to find out what is on your students' minds.

- While using chat has several advantages, it can be distracting for both you and your students.

- If you plan to use chat, establish some norms and communicate them to your students.

Chapter 6 – Work in Groups

The positive effects of working in groups has been well documented in the teaching and learning literature.[15] In an online live class with more than a dozen students, dividing the class into small groups can be a great way to get participants to engage, do some work, reflect, and learn from their peers. By splitting the class into small groups and giving students a meaningful task to complete or a meaningful question to answer, you give more of your students an opportunity to directly engage with the material, more time to process and make meaning of what has happened in class, and more chances to ask questions they might not feel comfortable asking in class. Working in groups also gives students an opportunity to learn from others who might be better able than you to understand the difficulties faced by someone learning the material for the first time. When you have a whole class discussion, the students who are not speaking might tune out more easily than when they are discussing in small groups.

Having students work in groups also has some drawbacks. You might feel like you are losing control over the class, and unless you clearly lay out what you would like students to do (see details below), students might not use the time as productively as you would like.

In Zoom, you send students to what are called "Breakout Rooms," where they can work for a period of time. Breakout rooms is a Zoom feature that allows you to split your Zoom meeting into separate sessions for each group. When the students are placed into breakout rooms, their Zoom interface is identical to the regular Zoom interface except that they can see only the students in their group. The participants list and chat become private to the students in that room. During this time, you can visit these rooms to listen in on the conversations, and you can bring them all back into the main room whenever you like.

To create and send students to breakout rooms, you need to click on the "Breakout Rooms" item in Zoom's main toolbar. See "Tech Tips" section below for instructions on how to add this item to your toolbar if it is not already there.

Figure 6.1 – Sending students into breakout rooms using Zoom's main toolbar

Many students view breakout rooms in a very positive light. In a recent student survey of Harvard Kennedy School students, over one third of respondents mentioned breakout rooms when asked what had been effective for their learning. One student commented that breakout rooms "create a tangible social pressure to participate." Another one indicated that "breakout sessions are really nice for getting to have more intimate conversations with peers, similar to how we would do in person." And in our online executive education programs at the Harvard Kennedy School, many participants mention breakout rooms as an important component of the program both for learning and for getting to know other participants.

As with other technological tools in online learning, the most important guideline in the use of breakout rooms is to be purposeful with what you want to achieve. Having a clear task that the groups are assigned to work on, a clear deliverable, and a specified timeframe helps tremendously. Several students reported that they were often in breakout rooms without a clear sense of what they were supposed to do and wasted a lot of time trying to figure that out. Others reported that they were much more likely to use breakout room time to catch up with each other socially when the task wasn't clear or if they didn't feel accountable for completing it. See sample instruction slide below.

Figure 6.2 – Sample slide to show students before they head to breakout rooms

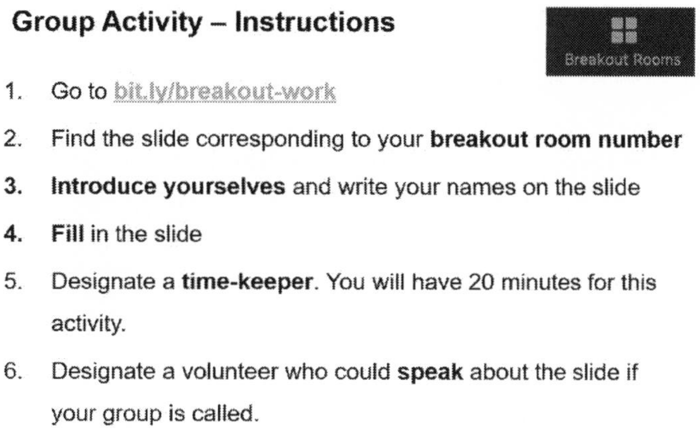

Group Activity – Instructions

1. Go to bit.ly/breakout-work
2. Find the slide corresponding to your **breakout room number**
3. **Introduce yourselves** and write your names on the slide
4. **Fill** in the slide
5. Designate a **time-keeper**. You will have 20 minutes for this activity.
6. Designate a volunteer who could **speak** about the slide if your group is called.

6.1 Main uses of breakout rooms

Breakout rooms can be used for many purposes. This section focuses on three key ones: (1) to answer a **question**, (2) to produce a **deliverable**, and (3) to build **community**.

<u>Use breakout rooms to answer a question</u>

If you are planning to ask a question that is central to the class you are teaching and where you think students would benefit from discussing and reflecting, breakout rooms might be a great tool for you. A typical pedagogic structure for doing this is within "Think-Pair-Share," a collaborative learning strategy widely used in K-12 classrooms.[16] The name comes from the sequence of the following activities:

- **Think:** You ask the students to think on their own about a particular question/issue.

- **Pair:** You ask each student to pair with a partner to discuss the question/issue.

- **Share:** You ask students to share a key insight from their paired discussion with the whole class.

This technique can be modified in various ways when teaching online. One of the most powerful ways is to conduct a poll after the "think" part. If the results of the poll indicate that the class is split, then you proceed with the "pair" part (See row A of Table 4.4 in chapter 4 for more details). This combination of polling with "Think-Pair-Share" is frequently referred to in higher education as "peer instruction," an approach popularized by Eric Mazur to teach physics and now widely used in many other disciplines.

In an online live class, the breakout rooms are used for the "pair" part of the sequence. You do not need to take the word "pair" literally, as groups can be larger than just two students. Slightly larger groups can be useful to minimize the chances of forming groups where everyone in the group answered the poll in the same way. See "How large should groups be" section below.

In Practice 6.1 – Using breakout rooms to answer a question

Rebecca Nesson, Associate Dean of the Harvard College Curriculum, teaches computer science and math courses at Harvard and uses breakout rooms to ask students to answer a series of challenging questions. At a recent class, she spent a few minutes at the beginning of class explaining the material of the day, and then assigned students to work on three questions. The students had access to a PDF document with the questions and discussed them in their groups. Because the subject was math, and answering the questions involved writing formulas and equations, students used a variety of tools (including Zoom's whiteboard and the latex equation add-on to Google Docs). While the students worked in groups, Rebecca and her teaching assistants briefly visited the breakout rooms to monitor the discussion and to help facilitate the conversation by asking probing questions. After the breakout rooms ended and the students had returned to the main room, the class went over the answers to the three questions, and new questions emerged. It was a very productive discussion because students had already spent time in the breakout rooms thinking hard about the questions and trying to answer them.

Use breakout rooms to produce a deliverable

One potential use of breakout rooms is to ask students to produce a deliverable, which typically means a shared document, slide, graph, or sketch. Requiring each breakout group to produce something concrete focuses their attention on the task at hand, particularly if students know they might be called on to present their work. It also makes their learning visible (to you, themselves, and other students), which has many pedagogical benefits.[17]

In Practice 6.2 – Using breakout rooms to produce a deliverable

Karen Brennan, who teaches several courses on education and technology at the Harvard Graduate School of Education, uses breakout rooms in very interesting ways. But before I tell you what she does, let me set the context for you. Observing Karen teach is like observing a concert where beautiful music is being played but the conductor is rarely seen. The students seem to be doing all the work in class. The casual observer might wonder how the class is going so well given that Karen seems to be doing so little! But what you don't realize until you speak with her and her students is how much work she and her teaching team have put in before class, advising and guiding the students individually and in small groups so that the class can go so well. It just seems like magic!

At a recent class, Karen started the session by asking a group of students to conduct a presentation about the topic of the day and explaining the activity that the whole class would be asked to do during the breakout rooms. Karen and her teaching team had assigned and briefed this group ahead of time. After this group finished its presentation, Karen assigned all students to breakout rooms based on their responses to a pre-class survey, matching students with similar interests related to that day's topic. In the breakout rooms, the groups focused on a clear task and documented their work on Google Slides (see below if you are interested in learning how to do this). The students in the group that had presented at the beginning of class were distributed across the breakout rooms and acted as facilitators of these discussions. After the breakout groups ended, each group was asked to present the slides they had produced in the breakout room to the whole class—in 2 minutes or less.

Use breakout rooms to build community

Another potential use of breakout rooms is simply to build community. This tends to happen anyway when you use breakout rooms for other purposes, but sometimes you might want to use them primarily for community building. See chapter 11 for an example.

6.2   Implementing breakout rooms

A number of questions arise when contemplating the use of breakout rooms during a live online class. Below are a few of the key questions with suggested answers.

<u>How do I assign students to groups?</u>

There are two ways to assign students to groups: randomly or deliberately. If you assign deliberately, you can do so before class or during class.

Figure 6.3 – Ways of assigning students to breakout rooms

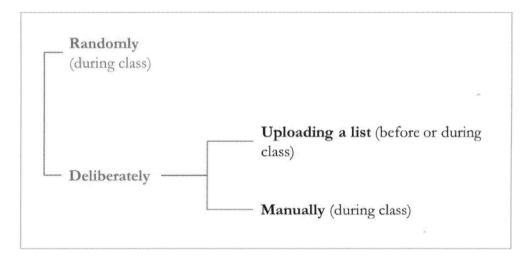

Assigning randomly is the simplest. It requires no planning. When you are ready to assign the groups in class, you click on "breakout rooms" in Zoom's main bar and then select "automatically" when prompted, indicate how many groups you want to have, and that's it. Another advantage of assigning randomly is that you might get students who would normally not interact with each other to work together. And if you use breakout rooms frequently during your course, assigning randomly might help students get to know many other students in the class (in contrast to the typical experience in a physical classroom, where students tend to socialize with other students they know or students who are sitting next to them).

Figure 6.4 – Assigning randomly to breakout rooms during class

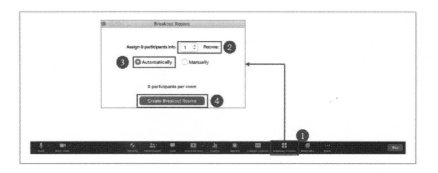

Assigning deliberately requires a bit more work. If you want to assign deliberately to breakout rooms, there are two ways of doing so: uploading a list of students with their respective groups (before or during class) or assigning students manually (during class). As explained in the "Tech Tips" section below, I recommend against uploading the list and in favor of assigning manually. To assign students to breakout rooms manually during class, see figure below. Assigning manually can be time-consuming, especially if you have many students. See "Tech Tips" section below for some possible ways to address this challenge.

Figure 6.5 – Assigning manually to breakout rooms during class

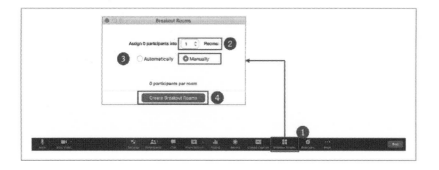

There are several reasons why you might want to assign students deliberately to breakout rooms. For example, you might want to have groups that are working on a project for the course meet during one of the breakout room activities in class. Or you might want to assign students to groups according to their interests or the activities they want to pursue. To do this, you can conduct a non-anonymous poll (before or during class) and then assign students according to their poll responses (see example slide below for how this can be done during class).

Figure 6.6 – Assigning students to groups according to poll response

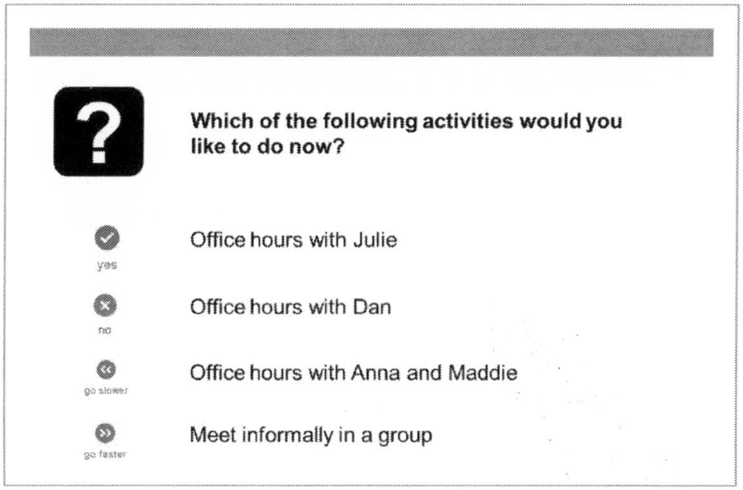

Independently of whether you assign students randomly or deliberately, once the assignment is completed, you will get a list of which students were assigned to which group, and you will be able to rename groups, reassign individual students to other groups, and tweak some settings (see figure below). Once you are ready, you click on "Open All Rooms" and students are immediately sent to their respective breakout rooms.

Figure 6.7 – Screen that shows up after you assign students to groups

How many students should I assign per group?

For most activities, four to six students per group is ideal. If you use breakout rooms for answering a question (rather than for producing a deliverable), groups of two to three students are often fine, especially if the goal is to quickly discuss the question. In general, groups larger than six students are not ideal if part of your goal is to have every student in the group actively contribute. Also keep in mind that the larger the group, the more time you probably want to allow for breakout rooms as it's often harder to coordinate larger groups. Finally, you can have a maximum number of breakout rooms (somewhere between twenty and fifty, depending on the number of participants), so if you have a very large class, take this into account in deciding group size for breakout rooms.

<u>How long should the breakout room activities be?</u>

One of the main concerns that students report about the use of breakout rooms is that they are not given enough time to complete the assigned task. So my advice is to err on the side of giving too much time rather than too little, although I recognize that as instructors we rarely have the luxury of time. The main factors that drive how long breakout rooms should be is the nature of the task students are asked to complete, and the number of students in the group. If the breakout groups are small and their purpose is to answer a question, the breakout session might be as short as 5 minutes. If it is to produce a deliverable, the 10-20 minutes range is more appropriate. I have seen some instructors dedicate as long as 30 minutes to breakout rooms, which can work if the students in the group are working towards making progress on an important task for the class or course.

Another factor to take into account in deciding time for breakout rooms is whether students in the groups are likely to know each other. If they are, they can probably start to work quickly. But if they have never met, consider allowing for a little time to introduce themselves and perhaps discuss an ice-breaking question, as described in chapter 11 on community building.

Finally, regardless of how much time you assign to the breakout room activity, there are three things to keep in mind. First, you can set the time you want the breakout rooms to last before sending students to the breakout rooms by tweaking the settings (see "Tech Tips" section below). Second, you can adjust the time designated for breakout rooms during class if you sense that students are done before time is up or if you think they need more time. Third, it is probably a good idea to give students a 3-minute or 5-minute warning before the end time (through the use of broadcast messaging; see "Tech Tips" section below), so they know they have to bring the discussion to a close.

<u>What should I do *while* students are in breakout rooms?</u>

There are several possibilities, including visiting one or more groups to check in (by clicking on "Join" next to the group name), checking the work they are producing if you asked them to document it in a place you can check (e.g., a Google Slide; see "In Practice" examples below), planning what you will do in class after the breakout room is over, or simply taking a break!

What should I do *after* the breakout rooms end?

In general, it is a good idea to devote some time after the breakout rooms to build on what the students did during them. A typical activity is to invite some groups (or individual students) to answer the question that was posed or present the work that they produced. When you conduct these activities, make sure that they are adding value beyond what students did in the breakout rooms. You probably don't want to simply ask all groups to report out their answer to the same question, as this may not be a good use of time. One possible approach is to build on their answer to address a bigger question – or perhaps the next question – in your class plan. For example, if they did some calculation during the breakout room, you can have one student quickly go over the calculation and then ask what implications that number has for a decision, or what were the key assumptions underlying the calculations.

What tools should students use to produce a deliverable during breakout rooms?

The answer to this question depends on a number of factors, especially what task(s) you want them to perform. There are a number of collaborative tools on the web to do different kinds of tasks, but one strong contender is the Google suite (particularly Google Slides and Google Docs). One way to use Google Slides during breakout rooms is to have a Google Slide document for all students in the class, and assign each group to one slide (or one set of slides) in the document. To do this, you can create a template slide(s) for one group and reproduce it for every group in your class.

Using Google Slides in this manner has the following key advantages: (1) Most students are familiar with the tool, so they don't have to spend part of their breakout room time figuring out how to use it; and (2) You can observe what students are writing in real time. Being able to monitor the groups' progress in real time allows you to visit groups that seem to be making little progress (or send a teaching assistant to do so), and to start planning whom to call on once the students come back from the breakout rooms to the main session. When you call on a group, you can display their slide (by sharing your screen with the students; see chapter 8) to provide this visual support when they present to the whole class. You can then call another group with a different, contrasting, or complementing view, and generate an interesting debate or discussion. If you are interested in the use of Google Slides for breakout rooms, see the section "In Practice" below for some examples, and the "Tech Tips" section further below for a step-by-step guide on how to implement this approach.

In Practice 6.3 – Using Google Slides in breakout rooms

Julie Wilson, a sociologist who teaches at the Harvard Kennedy School, uses breakout rooms to help prepare groups to contribute to case discussions. In a recent class where she was teaching about a social program in Jamaica, she wanted the students to understand the potential points of failure in the implementation of the program. She assigned different breakout rooms to represent each of the stakeholders involved in implementing the program (central government, local office, schools, etc.) and had each group fill in a table that identified flows of information, people, and money that involved the specific stakeholder they were assigned to. See sample slides below and the companion site for access to the full presentation.

What is most impressive about Julie's use of Google Slides is what she is able to do with what the students are writing during the breakout sessions. She monitors the slides as they are being written and identifies which group(s) she wants to participate in the post-breakout room discussion, based on what they are writing on the slide and on whether the group contains any students who haven't participated much in the course. When students come back from the breakout rooms to the main room, she meticulously proceeds to call on the groups that she selected during her examination of the Google Slides, and nudges the low-participation students to contribute to the conversation. In this way, she is able to conduct a discussion that is both effective and inclusive.

Figure 6.8 – Sample slides used by Julie Wilson for breakout rooms

Breakout Room 1 - Central Government

Group members:

Step of the process	Information	People	Money

Group spokesperson:	Group Timekeeper:

Breakout Room 2 - Local Office

Group members:

Step of the process	Information	People	Money

Group spokesperson:	Group Timekeeper:

Note: Each breakout room has its own slide (in the same Google Slide deck); only two are shown here.
See <u>companion site</u> for details.

 **Tech Tips**

<u>What do I do if I don't see "Breakout Rooms" in my Zoom main toolbar?</u>

If the "Breakout Rooms" feature does not appear on your Zoom bar, you need to add it. To do so, follow these steps:

- Login to your Zoom account using your browser. The web address (URL) will depend on your institutions' setting.

- On the left navigation bar, click on Settings. See image below.

- Click on sub-menu "In Meeting (Advanced)" and scroll down until you find the item related to breakout rooms. Make sure it is turned on.

How to assign deliberately to breakout rooms?

As indicated above, there are two ways of assigning deliberately to breakout rooms. First, you can upload a list of students with their respective groups before or during class. In practice, this method does not work so well for several technical reasons. If students register with an email address different than the one you have in the list you uploaded, they won't be sent to the breakout room and they will be left in the main room. Furthermore, sometimes students log in with more than one device to the same session (e.g., if their laptop camera does not work, they might log on with their laptop to engage in the session and view the material, and with their smartphone to provide their video feed), and this causes problems when pre-assigning them to breakout rooms.

The second way of assigning students deliberately to groups is to do so manually during class. If you have a small class, this is easy to do. But if you have a large class, this can be time-consuming. Two possible solutions you might consider:

- It may be helpful if you can designate someone (e.g., a colleague or a teaching assistant) to do it for you. The way this would work is that once the session starts, that person can create the breakout rooms with the students who are in class and according to the list that you gave them, and once you are ready to assign the breakout rooms, you ask this person to send the students to their breakout rooms. This is more effective, because your colleague can do this while you are conducting class, so the breakout rooms are ready to go when you decide it is time for this activity.

- Ask your students to rename themselves according to their group number (e.g., "1 – Maria Klinger," "2 – Ian Sullivan," etc.). Since Zoom arranges the participants list in alphabetical order, students from group 1 will appear at the top, followed by students from group 2, and so on, which makes assigning to breakout rooms much quicker since the students are already in the right order. This tip is courtesy of Horace Ling. For instructions on how to rename on Zoom, please see "Tech Tips" in chapter 10.

I hope that one day Zoom will make it possible to have students self-select into the groups they want and/or allow the instructor to send students to different breakout rooms according to how they answered a poll question.

Who should assign students to breakout rooms?

Only the host can assign to breakout rooms. This means that if you want another person (e.g., a teaching assistant, colleague, etc.) to assign and send students to breakout rooms, this person needs to be the host of the meeting and assign you as a co-host. If this is not your default setup, these roles can be reassigned at the beginning of class. The host is also the person who can broadcast messages to all rooms, and who can send people (you, a teaching assistant, etc.) to individual breakout rooms. You can visit several breakout rooms, but only one at a time.

How do I tweak the settings in breakout rooms?

Once you assign students to breakout rooms, Zoom will show you a screen that indicates how the students were distributed into the breakout rooms. If you click on Options, you can tweak a number of settings. One that I recommend you check is the one that indicates "Move all participants into breakout rooms automatically," as this means that whenever you send students into breakout rooms, they don't need to do anything to get there. Otherwise, students need to actively approve, which often delays the process. The other settings are mostly a matter of personal preference, but I think leaving the countdown timer at 60 seconds is probably a good idea as it implies that students will receive a warning that 60 seconds later they will be automatically returned to the main room.

Figure 6.9 – Tweaking the settings in breakout rooms

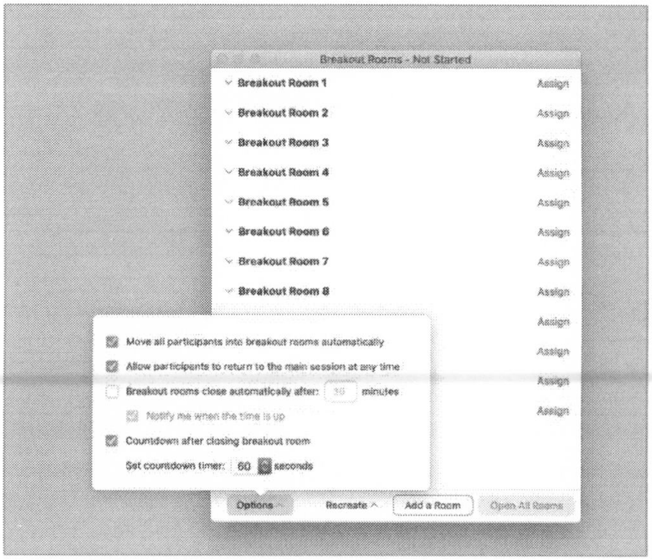

How do I broadcast a message to students while they are in breakout rooms?

When students are in breakout rooms, your breakout room dialog box (click on breakout rooms in the Zoom main bar if you don't see this dialog box) will give you the option to broadcast a message. See figure below.

Figure 6.10 – Broadcasting a message to breakout rooms

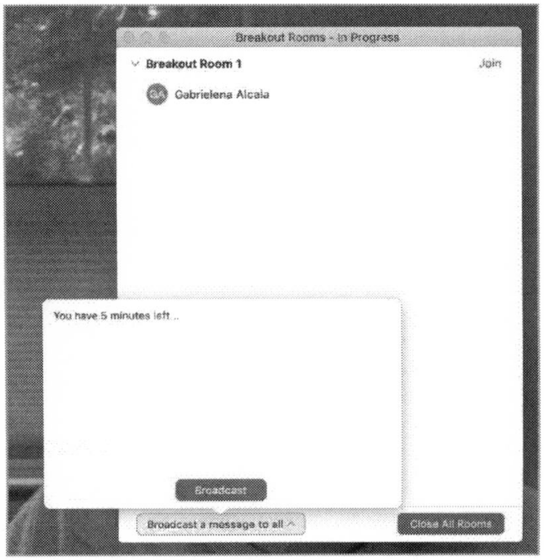

How do I end breakout rooms?

When students are in breakout rooms, your breakout room dialog box (click on breakout rooms in the Zoom main bar if you don't see this dialog box) will give you the option to "Close All Rooms." See picture above. Once you click on it, all students will be brought back to the main room after 60 seconds (or some other specified time if you tweaked the settings; see "Tweaking the settings in breakout rooms" above)

How do I produce and use Google Slides for breakout rooms?

See sample checklist below.

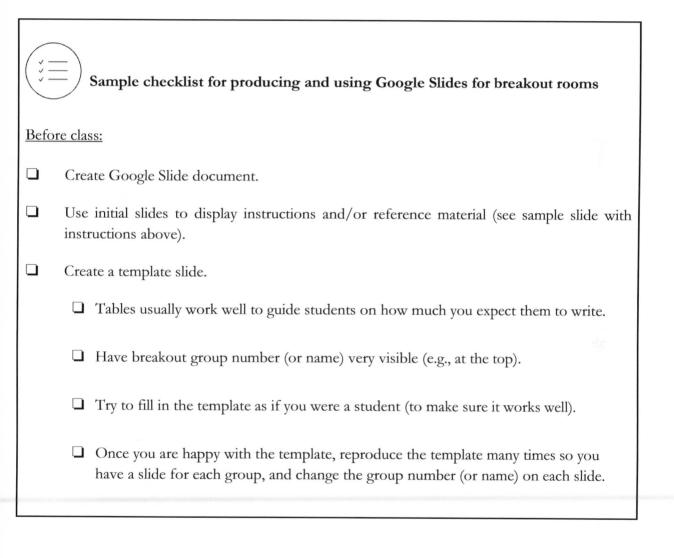

Sample checklist for producing and using Google Slides for breakout rooms

Before class:

❑ Create Google Slide document.

❑ Use initial slides to display instructions and/or reference material (see sample slide with instructions above).

❑ Create a template slide.

    ❑ Tables usually work well to guide students on how much you expect them to write.

    ❑ Have breakout group number (or name) very visible (e.g., at the top).

    ❑ Try to fill in the template as if you were a student (to make sure it works well).

    ❑ Once you are happy with the template, reproduce the template many times so you have a slide for each group, and change the group number (or name) on each slide.

❏ Set permissions in Google so that your students can edit the slides. If you forget to do this, your students will be able to view the slides when they click on the link, but won't be able to edit their group slide(s)!

❏ [Optional] Get a short link (though bit.ly, tinyurl, or a similar service) so you can more easily share link with the students.

❏ Technical tip: As of the time of writing, Google Docs and Slides do not work very well when more than a certain number of people are editing the document at the same time. I was able to do this well with eleven groups of five students, but it broke down with twenty groups of five. In this latter case, using two separate Google Slide decks (one for groups 1-10, and another for groups 11-20) will do the trick (tip courtesy of Teddy Svoronos).

<u>During class</u> (before sending students to breakout rooms):

❏ Go over the instruction slide.

❏ Give them the link to the Google Slide document.

❏ Show them the slide(s) you want them to fill in and explain the task as clearly as possible.

❏ Ask if there are any questions.

❏ Send (or have host send) students to breakout rooms.

Chapter Summary

- There are a lot of benefits to having students work in small groups during part of your online live classes.

- For this to work well, being explicit about the task you want students to accomplish (e.g., answer a question, produce a deliverable, etc.) and the time they have available is crucial.

- Consider having students document the work they do in breakout rooms (through Google Slides, Google Docs, or other collaborative documents) to make learning visible to you, themselves, and their classmates.

- Following the breakout room with a debrief or discussion is highly recommended.

Chapter 7 – Share Work

Another way students can engage in a live online class is by sharing their work with you and their peers. Sharing their work with the class can have many benefits for your students, including increasing their level of engagement, fostering a greater sense of accountability, making their work more visible, developing a greater sense of ownership over what happens in class, and building a greater sense of a learning community.

Students sharing their work also has some potential drawbacks. It takes time away from other activities in your live class, it might make you feel like you are losing some control over what is happening in your class, and the experience might end up not being as good as you want for the student(s) presenting (especially if they have not prepared adequately).

So like everything in teaching (and life), you will want to weigh the potential advantages against the potential disadvantages when deciding whether to offer students opportunities to share their work with the rest of the class. My sense, though, is that most of us employ this teaching approach too rarely rather than too often, so I will err on the side of nudging you to consider doing it more often.

The rest of this chapter describes how students can share their work and a handful of situations in which you might consider them doing so.

7.1   How students can share their work

Students can share their work on Zoom in several ways, including:

- By clicking on "Share Screen" in their Zoom's main toolbar and sharing the application or content they want to show. If you are unfamiliar with the use of "Share Screen" to share content, please refer to chapter 8.

- By having you click on "Share Screen" in your Zoom's main toolbar and share the content that your student wants to share. The students would send you the content in advance (e.g., by emailing you their PowerPoint presentation) or produce it in a shared space (e.g., Google Doc, Google Slide, Dropbox file, etc.).

- By showing something (sheet of paper, object, artwork) to the class through their Zoom video feed. In this scenario, it would be best for you and your students to use the "Speaker View" (instead of the "Gallery View") in Zoom.

- By performing (e.g., singing, acting, playing an instrument, etc.) "in front of the class" through their Zoom video feed. In this scenario, "Speaker View" would also be recommended.

- By allowing you or other students to access their computer through Zoom's Remote Control feature.

Which of these ways is optimal for your students to share their work will depend on what they want to share and the specific context.

7.2   Situations for students to share their work

There are many situations in which students can share their work in a live online session. This chapter focuses on the following:

- Share work they did before class

- Share work they did individually during class

- Share work they did in groups during class (breakout rooms)

- Share work in special events

- Share work that requires close examination

Share work they did before class

When you assign students to work outside of class on material that you think could be useful to examine in your live class, consider asking them to present this work in class. For example, if you have asked students to summarize their key takeaways from the previous class or how they would apply or extend the lessons from last class, you could begin the class by asking one or two students (or groups) to present their work in the first 3-5 minutes of class. Another possibility is for you to ask a student to elaborate upon a particularly interesting comment they made prior to class (e.g., on the discussion board of your learning management system), which you could reproduce in one of your slides. The work to be shared does not necessarily have to come in the form of presentation slides. For example, your students might show a physical object (e.g., a rocket ship) they built, go through a spreadsheet in which they recorded data from an experiment they conducted, or play some music they created. Whatever it is that you asked them to do outside of class, think about whether there are ways they can share it (or an aspect of it) during the live class.

One advantage of asking your students to share work they did before class is that knowing they might be asked to present publicly in front of their peers might stimulate them to exert greater effort and creativity than the traditional assignment submission where you are the only one who sees their work.

One question that arises if you ask students to present work they did before class is whether you should warn ahead of time the specific student(s) you selected to present. On the one hand, if you don't announce which students will be in charge of presenting, it might encourage *all* your students to prepare and exert a greater effort in the work they do outside of class. On the other hand, warning ahead of time the students who will present will likely lead to better preparation and a greater chance of a successful presentation.

Share work they did individually during class

At various points in your live class, you could stop the class, ask the students to do some work individually, and then when you all come back together as a class you could ask a student to present their work. For example, in a math class, students could present how they solved a problem. In a language class, students could read out loud a paragraph they wrote in response to your prompt. More generally, whatever you ask students to produce during the time of individual work, think about ways they could share this with the rest of the class other than describing it verbally. See "In Practice" session below for an example.

In Practice 7.1 – Using interactive verbal roll to create community

Kathy Pham is a computer scientist whose work has spanned Google, IBM, and the federal government at the United States Digital Service at the White House, where she was a founding product and engineering member. She teaches a course on Product Management and Society at the Harvard Kennedy School. At a recent online live class, she paused the class and asked her students to spend a few minutes individually developing a prototype of a food delivery app. She gave the students a few minutes to do this with pen and paper, and then called on a few students to present what they had come up with. The students presented their work by showing their sheet of paper and pointing to the various app "screens" that they had developed. It was a very simple way of sharing and led to a very interesting discussion about principles of product design. See picture below.

Figure 7.1 – A student shows the rest of the class a prototype of the app she designed

Note: Student in the picture gave permission to publish this image

Share work they did in breakout rooms

As indicated in the previous chapter, breakout rooms work best when students have a clear task at hand, and they feel that doing this task well makes a difference. One way to accomplish this is to have some students present the work they did during the breakout rooms. If they produced a deliverable (e.g., Google Slide), they could present it during class, and you could probe on some of the aspects of the work that you think are helpful to advance your learning goals for that live class session. See chapter 6 for more details on the use of breakout rooms.

Share work in a special event

Some special events traditionally done in person are now being done in live online sessions. For example, in the context of a class or academic program, you might have a celebratory event (e.g., commencement, graduation, etc.) happening online. One way in which the event could become more meaningful to participants is if they can share something about their overall experience (e.g., what it meant to them, their key lessons learned, etc.). See "In Practice" below for an example.

In Practice 7.2 – Allowing students to share in a special event

Anna Shanley and Maddie Meister work in executive education at the Harvard Kennedy School. In the closing session for one of the online programs they recently directed, they asked participants to produce one slide per group indicating one way in which their thinking had changed as a result of the program and one thing they were grateful for. They had shared this slide before the closing session using Google Slides, which made the logistics much easier. During the live online ceremony, each group took one minute to present their slide before receiving their (virtual) certificate of completion. By participating in this way, I think the ceremony was more personal and meaningful for the participants than the typical in-person experience. See sample slide below.

Figure 7.2 – Sample slide for certificate ceremony of executive education program at the Harvard Kennedy School

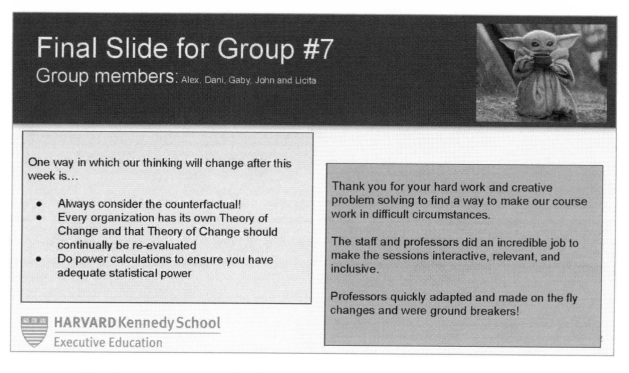

Note: Group member names have been changed to protect confidentiality of participants.

Share work that requires close examination

Sometimes students produce work that requires closer examination than what is possible to do by simply seeing a slide or object being presented. In these cases, you might benefit from the ability to see and change what is in the computer of one of your students. For example, if a student wrote a computer program, they might benefit from seeing how you make changes to improve or debug their code. Zoom's "Remote Control" feature allows one participant to take over control of another participant's computer. They both have to consent, and they can stop sharing at any time. See "In Practice" session below for an example. The companion site has instructions for how to use the "Remote Control" feature in Zoom.

In Practice 7.3 – Using Remote-Screen to have students share their work

At a Liberal Arts Remote Teaching webinar in March 2020, Ella Foster-Molina (Swarthmore College) and Jingchen (Monika) Hu (Vassar College) described how they have used remote sharing in their teaching. Hu uses it to help students debug their computer code in her data science course. One of her students gives her access to their computer and she can explore the student's code, fix a portion of the code by typing new code, and run the code again. See picture below. Foster-Molina also uses remote learning and thinks there are a number of circumstances where it simulates in-person interactions very well, but warns that "it should only be done if all of the parties involved are completely comfortable with a variety of privacy, autonomy, and safety concerns with looking at somebody else's screen and being able to actually control it." The companion site has a link to the webinar where they presented this work.

Figure 7.3 – Jingchen (Monika) Hu (Vassar College) uses Remote Control to help debug computer code produced by her students

Chapter Summary

- There are many benefits to having your students share their work in front of you and their peers.

- These benefits will of course need to be weighed against the benefits of doing other things in the live class. Time is limited.

- Your students can share work in a variety of situations, including:

  - Share work they did before class
  - Share work they did individually during class
  - Share work they did in groups during class (in breakout rooms)
  - Share work in special events
  - Share work that requires close examination

- Whenever you ask students to produce work, think about the potential benefits of having some of them present this work in a live class session and the ways in which this can be done most effectively.

PART III – WAYS YOU ENGAGE

This part describes how you, as the instructor, can engage in a live session beyond the ways implied in part II, which focused on how students themselves could engage (by speaking, voting, writing, working in groups, and sharing work). Chapter 8 assumes you have content (such as slides, written material, videos, or websites) that you want to present to your students. It explains various ways you can present this "preconstructed" material. Chapter 9 assumes that you would like to "construct" during the live class session part (or all) of the material you present, in the form of annotating pre-existing materials (e.g., by writing or typing on a slide) or writing or drawing from scratch (e.g., by writing on a physical or electronic whiteboard). Since chapter 9 builds on chapter 8, I suggest reading chapter 8 first.

Chapter	You can:	Key Zoom tool(s)
8	Present	**Hardware**: Computer
		Software: Presentation software (e.g., PowerPoint, Keynote, Google Slides, etc.), Web browser (e.g., Chrome), Video Player
		Zoom Tools: Share screen, Zoom backgrounds
9	Annotate	**Hardware**: Computer, tablet, document camera, physical blackboard, flipchart
		Software: Presentation software (e.g., PowerPoint, Keynote, Google Slides, etc.)
		Zoom Tools: Share screen, Whiteboard

Chapter 8 – Present

During a live online session, you may want to present to your students material that you prepared prior to class, such as slides, written material, videos, and websites. This chapter explains how to present this kind of "pre-constructed" material on Zoom, with a particular emphasis on slides.

You present material in Zoom by sharing your screen using Zoom's main toolbar. Once you do so, you can share content from a number of sources, including:

- A specific application (e.g., PowerPoint, Google Chrome, etc.)

- Zoom's Whiteboard feature

- Video from a secondary camera (e.g., document camera, smartphone, etc.)

- iPhone/iPad screen

- Audio from your speaker

Zoom shows you the sources that it can share in a dialog box. Notice that for an application to show in Zoom's dialog box, it has to be open on your computer. So if you plan to present from an application (e.g., PowerPoint), make sure that this application is open before you click on "Share Screen" in Zoom's main toolbar.

Figure 8.1 – Presenting material through the "Share Screen" feature in Zoom's main toolbar

Figure 8.2 – Using Zoom's "Share Screen" dialog box to share an application on your computer

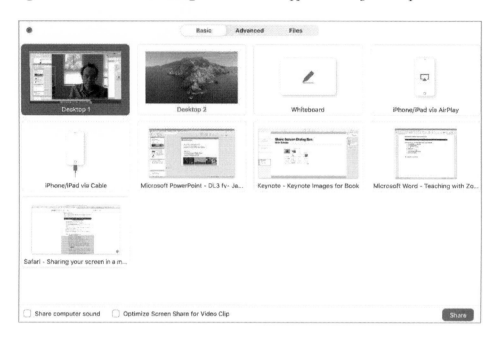

This chapter focuses on how to share your presentation slides, as this is the main type of pre-constructed material used by most instructors. The chapter ends with instructions on how to present two other types of materials: videos and internet browser pages.

Before we delve into how to present slides (or similar material), a word of warning which reinforces some of the pedagogic principles underlying this book: presenting slides for a long time without engaging directly with the students is unlikely to effectively stimulate their learning. This is true in a physical classroom and even more so in a virtual classroom, where you have little control over what students are doing while you engage in your presentation. Remember the mantra from Doyle[18]: "The one who does the work does the learning." When you present your slides, you are the one doing the work, so think about your presentation in the service of stimulating other activities in class that help your students learn and engage. Think about constantly engaging your students in doing some of the work. If this has not happened in your class for more than 5-10 minutes, it might be time to pause and get them to engage.

Finally, when presenting slides, remember to follow principles for preparing, designing, and delivering good presentations. These principles are beyond the scope of this book, but the <u>companion site</u> has a few useful references.

8.1   Presenting your slides

<u>How do I present my slides?</u>

If you simply want to show a presentation to your students, the easiest way to do this is to "Share Screen" in Zoom's main toolbar and then double click on the icon corresponding to your presentation software (e.g., PowerPoint, Keynote, etc.). Once you do so, Zoom will signal that you are sharing this application by placing a border on your presentation software. For all of this to work, make sure that your presentation software is open before you click on "Share Screen" (so that it appears as one of your options in Zoom's Share Screen dialog box). For reasons explained in the "Tech Tips" section below, I suggest you share your slides using the icon corresponding to your presentation software (e.g., PowerPoint) and not the icon corresponding to your desktop; this item might be called "Desktop" or "Screen" depending on the version of Zoom you have.

Figure 8.3 – Sharing a PowerPoint presentation with your students

Select PowerPoint icon in Share Screen.

Your PowerPoint "Window" will then be shared with your students.

After you share your screen, others in your Zoom session will see whatever is happening on that screen (e.g., presentation goes into slide show mode, you advance a slide, you write on a slide, etc.) until you click on "Stop Share".

Should I share my slides all the time?

When you share your slides, your students' screens change quite a bit to accommodate the slides. The table below shows how their screen typically changes; the particularities will depend on your students' preferences and device setup, but two things emerge: (1) Your slides occupy a lot of the students' screens, and (2) their attention is likely to deviate from you to the slides. In other words, the moment you share your slides, they are likely to take over a large share of your students' screen and attention. While this is also the case when you teach in a physical classroom, the change is more dramatic in a virtual classroom where your image is relegated to a small square on a screen dominated by your slides.

Figure 8.4 – How your students' screens change when you share your slide

Your students' screen before you share

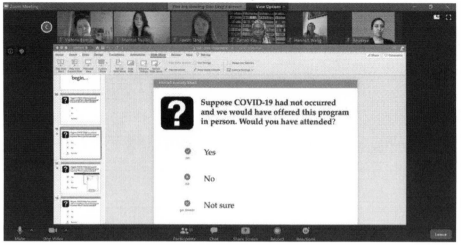

Your students' screen after you share

Given these facts, some faculty colleagues I have spoken with recommend a judicious use of slides. One of them advised using slides "only when absolutely necessary." Others have developed interesting ways to present material without having slides take over their students' screens. See "In Practice" below for some innovative solutions.

Finally, also note that when you share your presentation, it will take over a large part of your own screen. For this reason, some instructors like to use two monitors: one to look at their students' video feeds and other items (e.g., the participants list, the chat window, etc.), and the second one to look at the slides being presented. A second monitor occupies additional space on your desk and costs money, but if you plan to present slides or other material through Zoom and can get access to a second monitor, I highly recommend doing so. Preserving the ability to look at your students while you present enables you to see how they are responding to your presentation, including how attentive they are, and a second monitor makes it much easier to do this. The "Tech Tips" section at the end of this chapter has more details on the use of two monitors.

In Practice 8.1 – Presenting slides without sacrificing screen real estate: High tech edition

Mitch Weiss teaches entrepreneurship courses at the Harvard Business School and has a creative way of presenting slides without crowding his students' screens. For his case discussions, he uses two sets of PowerPoint slides. The first set is used in a traditional style where the slides occupy half of his students' screens. He uses Zoom's Share Screen feature to show them these slides, and annotates by typing on the slide (see next chapter). The second set of slides are ones that don't have as much information and are typically there to help set the context and have a conversation. When he is presenting these latter slides, he wants students to focus on him and each other more than the slides. So instead of sharing the screen, he sets the slide as a Zoom background (see "Tech Tips" section below for how to do this) and the resulting effect is similar to when a TV news anchor is broadcasting from a studio. He also uses this technique when he asks students to answer non-anonymous polls. See pictures below.

Figure 8.5 – Presenting slides as Zoom background

Mitch Weiss presents contextual slides as Zoom background

Mitch Weiss presents slide with poll instructions as Zoom background

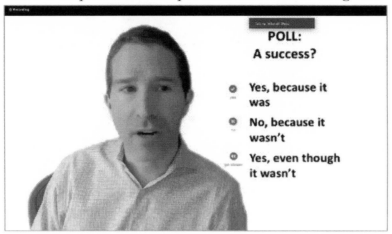

In Practice 8.2 – Presenting slides without sacrificing screen real estate: Low tech edition

Renee Loth, a distinguished journalist and opinion columnist for the Boston Globe, teaches communication courses at the Harvard Kennedy School. At a recent workshop, she was teaching a session titled "The ABC's of Op-Ed Writing" to a group of students. As she was launching her session, she brought into her video feed a simple prop that spelled out the ABC's of the op-ed form. She then started explaining each of the three key ingredients. Since it was the only "slide" she presented in the workshop, the students had an easy way of associating her with the prop, and she was able to command their attention throughout the segment where she discussed each of the three key ingredients, creating a memorable learning experience. See picture below.

Figure 8.6 – Renee Loth uses prop to present material to her students

<u>Should I send my slides to the students before class?</u>

You might consider sending the slides (or making them available) to your students before class so they can have access to them during class. In this way, your students can print the slides or put them on a tablet to take notes during class, or simply display them on a second monitor if they have one available. Giving students access to the slides before class also means that they can refer to a slide that you have previously shown if they want some clarification or need to remember something, or to a slide you have not yet presented if they want to look ahead to what's coming. Because of this, if you decide to send the slides to your students before class, you probably want to eliminate from the version you send them any slide that is meant to provide an element of surprise or an answer to a question that you want students to engage with.

One concern with students having the slides during your session is that they might get distracted. I can see this point, but there are many other things they can get distracted with in an online learning environment. I think a well-designed set of slides (with space for note-taking) has more potential to help students focus than to distract them. For this reason, I favor making slides available to students before class, although I recognize that this is a personal preference.

8.2   Sharing video

Sometimes you might want students to view a video to amplify a class lesson. Just as in a physical classroom, when deciding whether to play a video or not, consider your pedagogic goals. Is this a video that students could watch before or after class instead of during class? If you are considering showing it during class, what makes this video an activity that benefits from everyone watching together at the same time? For example, perhaps you want students to share their immediate reaction, or the video is best understood only after you have taught some lessons in class and it sets up the class to learn a different lesson. If you can't think of a key reason to display the video in class, you can often save valuable class time by having students view it before or after class. See chapter 10 for more details on how to decide what parts of your content can be reserved for outside of class.

To share a video on Zoom, you follow a similar process to sharing a slide. You first select "Share Screen" from Zoom's main toolbar. Then you select the source you want to share from. This could, for example, be a video player (e.g., Quicktime) playing a file you have on your computer or a video from the internet (e.g., YouTube) that you want to play from your internet browser. A key difference when sharing a video that includes audio content (as opposed to sharing slides) is that you also need to click on some settings to optimize video and sound sharing (see picture below), and then click share. I recommend opening the application from which you plan to play the video and having it ready to go (i.e., at the exact point you want the video to start playing) before class. This enables you to start playing the video quickly during class. It is also a good idea to ask students whether they can see and hear well.

Figure 8.7 – Sharing a video that will play from an internet browser

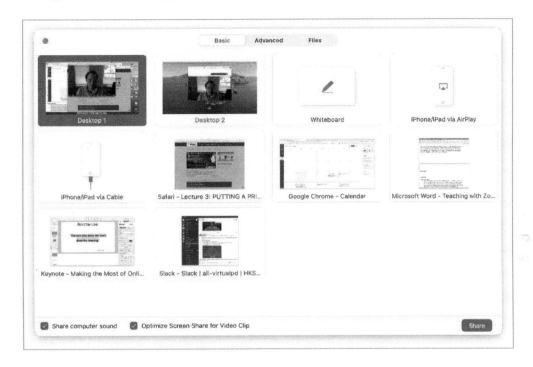

If you are worried about the quality of the video or sound degrading when transmitted over the Zoom meeting, an alternative is to type into the chat a link to a video file that students can play on their computer, and then resume class discussion after the elapsed video time has passed and students have watched it.

8.3   Sharing resources from an internet browser

Sometimes you may simply want to share resources available on your internet browser (a website, a video, an audio clip, etc.). This is especially useful if you want to visually show how to do something on the web or if you want to present work that students have done using collaborative software (e.g., Google Slides, Google Docs, etc.) as described in chapter 6 ("Work in groups"). One advantage of presenting from a browser is that you can switch what you are presenting easily by simply navigating from one browser tab to another. To share from an internet browser using Zoom, you can use the same process to share a video described in the section above.

 **Tech Tips**

Sharing your desktop

- When you press "Share Screen" you will have the option to share your computer desktop. It is the first item you see in Zoom's Share Screen dialog box (it might be called "Desktop" or "Screen" depending on the version of Zoom you have). Doing so means that Zoom will share whatever is on your computer desktop while you are sharing. This has the advantage that if you are planning to present from two or more applications (e.g., PowerPoint and your internet browser), you can quickly change what students are seeing by simply switching the application that is active on your screen. Otherwise, you need to stop sharing one application (e.g., PowerPoint) and start sharing the other one (e.g., your internet browser). You can also display the applications side by side (though beware that your students might have a small monitor, which will make it hard for them to read these well).

- The key disadvantage of sharing your computer desktop is that you might forget that you are sharing, and inadvertently share things that you are not intending to share. For example, I have heard colleagues tell stories of inadvertently sharing a personal email that they were quickly checking, a list of students that they wanted to call on, etc. For this reason, my recommendation is not to share your computer desktop unless you have very strong reasons to do so, and you trust that you can remember not to open on your desktop any files that are not meant for students to see. Instead, as described above, share the application you want students to see (e.g., Powerpoint, Keynote, etc.). My other recommendation is that if at any point during your Zoom class you are unsure of what is being shared, stop sharing and then share again.

Using two monitors

- To optimally use two monitors in Zoom, you need to "enable dual monitors" in the Zoom settings. The companion site has a link that explains how to do this for both Windows and Mac computers. Enabling dual monitors allows for gallery view, speaker view, shared screen, chat, and participants list to show in different windows. This feels like a much better experience than having a second monitor for more room without dual monitors enabled.

- If you are using two monitors and share PowerPoint from one of your desktops, the default is for PowerPoint to take over your two monitors (one in presenter mode, which has your slides and notes, and the other one in slide show mode, which displays the slides that students see). This implies no space to see your students. I think there are two ways to deal with this problem:

  o First, you can present PowerPoint from a tablet (e.g., iPad, Microsoft Surface, etc.), by connecting the tablet to your computer and sharing its output via Zoom's "Share Screen" feature.

  o Second, you can go to PowerPoint's "Slide Show" Menu, click on "Set up Show," change settings to "Browse by an individual (window)" and to advancing slides manually. See figure below and companion site for details. If you are using Keynote, use the new feature called "Play Slideshow in Window" in the Play menu to accomplish the same goal. If you use Google Slides, this problem does not exist.

Figure 8.8 – Changing settings in PowerPoint so that it does not take over the two monitors when you present

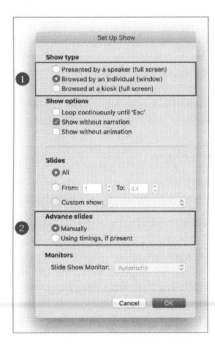

- Note that if you have two monitors, they will appear as Desktop 1 and Desktop 2 (or Screen 1 and Screen 2, depending on your version of Zoom) in your share dialog box. You could share either of them (though as I indicated above, I don't recommend sharing computer desktops).

- When you use two monitors, it can take a little while to get used to how the windows are distributed across the two monitors when you share and stop sharing your screen. Practice and try to customize your setup so that you are comfortable. This will minimize the need to move lots of windows while you are teaching.

Presenting using an external device

- If you plan to present using an additional device (e.g., iPad, Microsoft Surface, etc.), I suggest connecting that device to your computer by cable and then using Zoom's "Share Screen" feature, rather than using any of the other alternative methods (see below).

  - *Alternative 1:* Some people connect that additional device by logging into Zoom as a separate user on that device, and therefore have two users connected into Zoom (one from the main computer and one from the additional device). This means you have two devices using bandwidth on your internet network, which could lead to lower overall quality.
  - *Alternative 2:* Some people try to connect the iPad through AirPlay. I have found that this connection is not very stable when you have many students, and therefore recommend connecting the iPad through a physical cable to your main computer (and then sharing its screen by using Zoom's "Share Screen" feature and clicking on "iPhone/iPad via cable").

Using slides as Zoom background

- Update to this book on August 2020: In early August 2020, Zoom introduced a way to show slides behind you without using Zoom's original virtual background feature. The feature is called "Slides as Virtual Background" and can be accessed after clicking on "Share Screen" and then "Advanced". It is in Beta version at the moment of this book update, so I won't describe it in detail here, but it seems like a better solution than the one described in the bullet points below (which use Zoom's generic virtual background feature accessible in the Start/Stop Video icon on Zoom's main toolbar).

- Before explaining how to show a slide behind you in Zoom like Mitch Weiss does, a few words of warning:

  o Depending on the processing power of your computer, using a Zoom background might slow down your computer, and create other challenges for you.

  o For most people, having a virtual background creates some strange effects in the way your image is displayed on the video feed. In particular, your movement is sometimes not well captured when you have a virtual background. You can test this to see whether or not this is a problem for you.

  o While you can address some of these challenges with lighting, equipment (e.g., computer, green screen, etc.), and other things, be aware that for most people some of these issues are likely to persist.

  o This is not a reason not to use virtual backgrounds, but a warning to be aware of some of the limitations and a word of advice to use virtual backgrounds deliberately.

- Given the above, my advice is to experiment with one slide before investing too much time with this approach.

- In order to show a slide behind you in Zoom like Mitch Weiss does, the following steps are involved (see companion site for more details):

  o Create the slide(s) in your preferred software (e.g., PowerPoint, Keynote, etc.). When you do so, most of the slide should be left blank (remember that your face will be in front of most of the slide when you present it). Because this process involves some trial and error, follow the next two steps below for a single slide, and you can come back to create additional slides once you have the right template that works in your setting.

  o Save the slide as an image (e.g., PNG). How to do this depends on the software you are using. For example, in PowerPoint for Windows, select File > Save As and then select PNG from the "Save As Type" dropdown menu. In PowerPoint for Mac, select File > Export.

  o Join a Zoom Meeting, and add one or more slides to your list of virtual backgrounds by clicking on the arrow next to "Stop Video" on the main Zoom bar, selecting "Choose Virtual Background…", and the plus sign (+), and then find the PNG file on your computer. See image below. If you don't see the option of "Choose Virtual Background," it might mean that you need to enable it. See companion site for instructions on how to do this.

  o Once you are in class and want to select a slide as a virtual background, follow the previous step but instead of hitting the plus sign (+), click on the image corresponding to the slide that you want to display as your virtual background. If you plan to use several slides as your virtual background, you might want to leave the dialog box open so that you don't have to open it every time you need to switch your background slide.

Figure 8.9 – Adding a virtual background in Zoom

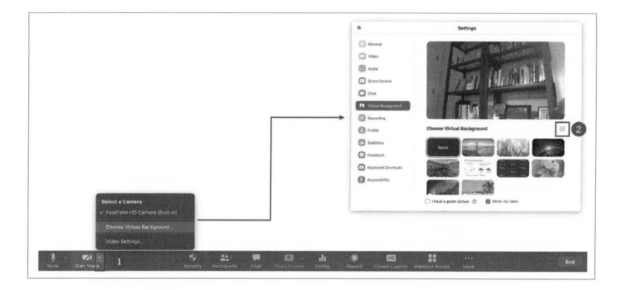

Chapter Summary

- You can use Zoom to present slides, videos, pages on your internet browser, and pretty much anything on your computer.

- Remember that presenting slides for a long time without having students actively engage in their learning is unlikely to be effective. Plan for interaction every few minutes of class.

- Be aware that the moment you share your slides, they are likely to take over a large share of your students' screen and attention. So do so judiciously, and consider not sharing your slides in segments of the class where you don't need them.

- When presenting, try to keep your eyes on your students rather than on your slides. If you have access to a second monitor, this goal becomes easier to achieve.

Chapter 9 – Annotate

If you use a blackboard, whiteboard, smartboard, document camera, or similar device when you teach in a physical classroom, you are probably used to the idea of annotating. I use the term "annotating" in a broad sense to include:

- Writing or drawing on a blank page/canvas/board

- Writing notes (by hand or typed) on an existing document (e.g., slide, text, diagram, etc.)

- Drawing a diagram, graph, or visual on an existing document

- Underlining, highlighting, or circling words on an existing document

Annotating to explain your ideas or record student comments can make your live online classes much more dynamic, engaging, and effective. In this chapter, I describe different ways to integrate the practice of annotating into your teaching – and then some ways of annotating with Zoom. As you look them over, I suggest you keep these guidelines in mind:

- **Before choosing any of these tools, test-drive a few of them.** Each has its own upsides and downsides, and it's important to choose a tool that works for you. For example, if you simply want a blank canvas to write on, using a physical flipchart or Zoom's native whiteboard tool might be enough, whereas if you want to write on your slides, using a tablet or document camera will be preferable.

- **Before using any of these tools with your students, try them out first on a friend, colleague, or family member.** Have someone (ideally more than one person using different types of devices) play the role of the student. Set up a Zoom meeting with them, and have them see and do exactly what your students would see and do. Try out all the tools you might use with your students. For example, if you plan to use a flipchart, have your friend in the Zoom meeting check that they can see the whole surface of the flipchart and can read your handwriting! The feedback you get will be priceless, and even small adjustments will make a big difference in your ability to use the tool successfully.

9.1　Different ways to integrate annotation into your teaching

You can use annotation to explain your ideas and/or to capture your students' ideas.

Annotating to explain ideas

A common goal of annotating is simply to explain or highlight ideas that you are presenting to the students. If you write on a board when you teach in a physical classroom, you are used to the idea of writing as a part of the process of explaining your ideas. Similarly, if you use a document camera or a smartboard, you are used to the idea of having a document or visual on which you can then write, draw, underline, and highlight.

Annotating to capture and make student comments visible

If you step into a typical case discussion in a classroom at the Harvard Business School (HBS), you will see the professor constantly interrogating the students on various aspects of the case, and writing the various student responses on one of the nine boards available in most HBS classrooms. The instructor breaks the discussion into different segments (called "pastures") and typically devotes a board (or two) to each pasture. In preparing for class, HBS faculty members are known to be meticulous in deciding what questions they will ask the students, what they will write on each board, how they will transition from one board to another, and how the boards should broadly look like at the end of class. The latter is called a "board plan" and it forms part of their teaching plans. It is very impressive.

While the majority of us don't plan how we will annotate student comments in class with this level of detail, there is something very powerful about being able to write down the essence of what students are saying – whether on a blackboard, a Google Slide, or on a tablet. Doing so conveys the notion that learning in your class is a collective enterprise and that students are expected to learn from you but also from each other. Students feel heard, are encouraged to contribute to class, and get more actively engaged in their education. They do not see themselves as passive receivers of knowledge, but rather as active creators of their own learning experience. Finally, when you write down the essence of what students are saying, you make their learning visible, and can refer back to some of what you wrote at various points in class to underscore or highlight key ideas.

As with every other aspect of teaching, being deliberate about what you want to accomplish is key. Whatever method you use for recording student comments, it might be helpful to prepare templates that set the structure for your note-taking. This is particularly important in the context of an online live class, where you will tend to be more constrained in terms of physical space. And you don't have to adopt a full HBS teaching style to benefit from this approach. You could decide that for a given segment of class, you will ask students a question (or set of questions) and write their answers in a place that is visible to the whole class.

9.2   Key question: Annotate by hand or by typing?

There are many hardware and software options to annotate during a Zoom class. A key decision that drives which annotating setup works best for you is whether you want to annotate by hand or by typing. While this is a matter of personal preference, the following guidelines might help you decide. Annotating by hand gives you more flexibility in terms of how you use the space, feels more natural to some instructors (and to some students), and you can more quickly produce a wider range of output (words, diagrams, underlining, highlighting, etc.). Annotating by typing typically leads to more legible and better-looking output, and does not require an additional device or piece of equipment. Finally, annotating by hand is not a great idea if your handwriting is hard to read, and annotating by typing is probably not a good idea if you type slowly like me. The table below summarizes the options I recommend you consider for each of the two ways of annotating.

Table 9.1 – Options to consider for annotating by hand vs. typing

Way of annotating	Key tools to annotate	Cost	Learning curve
By hand	Physical blackboard, whiteboard, or flipchart	$	Low
	Document camera to capture handwriting on paper	$$	Medium
	Smartphone to act as a document camera	-	Medium
	Tablet (e.g., iPad, Microsoft Surface, etc.)	$$$	High
Typing	Zoom's native whiteboard feature	-	Medium
	Presentation software (e.g., PowerPoint, Google Slides)	-	Low

Adapted with permission from document "Options for Virtual Board Work" produced by Ian Tosh and Kristin Sullivan for the Harvard Kennedy School. Cost column assumes that you don't already own the hardware (or have access to it), except for smartphone option.

Annotating by hand

There are many options for annotating by hand, including:

(1) Writing on a physical blackboard, whiteboard, or flipchart located behind you

(2) Writing on a sheet of paper on your desk and capturing output using a document camera, your smartphone, or a similar device

(3) Writing on a tablet

(1) Writing on a physical blackboard, whiteboard, or flipchart located behind you

If you are accustomed to using a blackboard or whiteboard to teach, you can continue doing so if you can adjust to what will likely be a more spatially constrained environment (like your home or office). I confess that I was skeptical about this approach being effective until I saw some instructors using it successfully. And if you can teach an online live session from a physical classroom, you might be able to use technology to provide your students with an experience that resembles being in a classroom. See "In Practice" examples below.

In Practice 9.1 – Annotating with a flipchart from home

Rem Koning teaches strategy courses at the Harvard Business School and uses a flipchart to annotate what his students say in response to his questions. His choice of using a flipchart is very deliberate. "The minute you share a slide in Zoom with the students, all their attention goes to the slide," he says. So he uses slides only occasionally. Most of the time what you see is him in front of a flipchart, alternating between asking questions and writing students' responses on the flipchart. Since the flipchart offers much less space than what he has available in a physical classroom, he is much more deliberate and constrained in what he annotates. He writes with a black marker but occasionally uses color markers to highlight key ideas. Once he finishes a segment of the discussion, he moves to the next segment and flips onto a blank page on the flipchart. In one session, Rem wanted to highlight the importance of the founder to a firm's success, so at that point in the discussion he taped a picture of the founder on the flipchart, which I believe made the point very memorable for his students. Rem is used to teaching in a classroom with nine boards, so if he can adjust to the space constraints of a flipchart, we all can! See pictures below.

Figure 9.1 – Rem Koning using a flipchart in a strategy class at Harvard Business School

Annotating in two columns of a flipchart

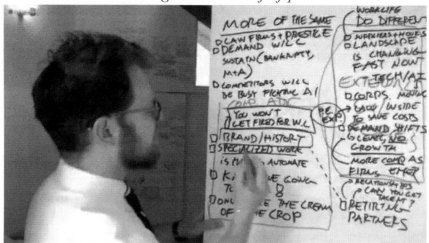

Taping a picture on top of a flipchart

In Practice 9.2 – Using a blackboard in a classroom

Steven J. Miller teaches mathematics at Williams College. He has taught courses where Williams College students are physically present with him in the classroom, while students from other liberal arts colleges (e.g., Swarthmore, Vassar, etc.) join in remotely. This is a common practice among small liberal arts colleges that want to expand the number of courses available to their students. In these classes, he uses technology that allows the camera to track him in the classroom while he writes on the board (if you are interested, see the <u>companion site</u> for an article describing the technology he uses). So far, he has mostly used asynchronous recordings of his classes to address two issues (students in colleges with different calendars, and students that want to take two courses on the same time block). He also considers recording as a valuable asset for students who have to miss class or for others who want to watch a portion of class again.

While you might anticipate being restricted to a small blackboard/whiteboard while teaching online from your home, Steven's example suggests that it might be feasible to use a full-sized blackboard in a live online class if you can get access to a classroom. At a webinar of liberal arts colleges, he commented on the use of a blackboard as one of the ways an instructor can annotate, noting, "I look at Zoom as a way to help facilitate conversations. Depending on what you are doing, different things will probably work better for you. Just be open to exploring." See picture below.

Figure 9.2 – Steven J. Miller teaches from a physical classroom and posts recording of class for students who cannot attend the live class

Source: https://www.youtube.com/watch?v=NgHIiZUYI6g&feature=youtu.be

(2) Writing on a sheet of paper on your desk and capturing output using a document camera, your smartphone, or similar device

If you love the feeling of writing with a pen on a piece of paper, you might consider doing so to capture your students' comments and sharing these comments with them using a document camera, smartphone, or similar device. One advantage of this setup is that you can use a blank page but also a printed page or slide, or even a page from a newspaper or magazine. You can then use a pen to write, highlight, annotate, underline, etc. This makes it a very flexible system. Two things to keep in mind as you consider this option are:

(1) It requires having space on your desk to write and to put the equipment.
(2) Anything you put under the camera will show up (i.e., you cannot be self-conscious about your hands!).

In Practice 9.3 – Annotating by using a document camera or similar device

Shoshanna Kostant, a teacher at Brookline High School in Massachusetts, uses a document camera to teach math. She sometimes has a solved-out math problem pre-written on a sheet of paper, and hides some of the content of that page by placing a blank sheet of paper on top of part of the original sheet. As she explains how to solve the problem, she progressively slides down her cover sheet revealing more of the content of the original sheet. This allows her to scaffold the learning for her students in a very simple way. Moreover, when she gets to a part of the problem that she wants her students to try solving, she pauses the class, gives her students some time to think, and then asks them to privately chat their answer. She then calls on a student and progressively continues sliding the cover down to make visible the solution to that part of the problem. See picture below. On other occasions, instead of having the math problem already solved, she just has the question pre-written and then writes the answer as a student reveals it verbally.

Figure 9.3 – Shoshanna Kostant scaffolding the learning of her students by sliding down cover page on top of a solved math problem

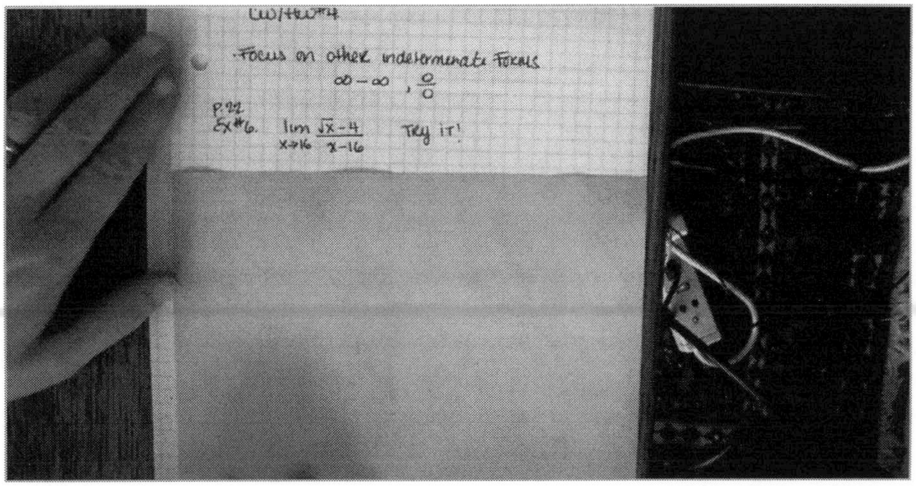

Figure 9.4 – Shoshanna Kostant's setup with a document camera

You can also use your smartphone as a document camera. For this to work, you need to place your phone in a way that it films the sheet of paper where you are writing. And you need to install an app on your phone that allows Zoom to recognize that your phone is acting as an additional camera in your computer. See "Tech Tips" section below for details. The advantage of using the document camera is that it is a device designed specifically for this use. The disadvantage is that it occupies space on your desk and is more expensive than using your existing smartphone (assuming you have one).

(3) Writing on a tablet

The most flexible option for annotating by hand is to write on a tablet. If you already use a tablet with an electronic pen (e.g., Apple pencil for an iPad, stylus for Microsoft Surface, etc.), this is probably an attractive option for you. The reason this method is so flexible is that you can essentially share anything on your tablet with the students. This means you can use whatever software you are comfortable with to write your students' comments. For example, both PowerPoint and Keynote allow you to handwrite on your slides while they are presented on a tablet, which provides you the benefits of modern presentation software along with the flexibility of being able to handwrite on your slides. You can also use software that is more specifically designed for note-taking, such as Microsoft OneNote, Notability, GoodNotes, and others. Finally, you can also use a tablet to write using Zoom's native whiteboard feature.

In Practice 9.4 – Using a tablet to annotate

Teddy Svoronos uses an iPad to teach quantitative methods at the Harvard Kennedy School. He imports a PDF of his class handout to an app called GoodNotes, which he uses to project the PDF and annotate it during class. As he asks questions to students and they reply, he annotates his handout with their answers. He also uses the pen to underline important parts of a table, highlight sections of the output he wants his students to pay special attention to, etc. The students are seeing their input reflected in how the handout is annotated, and he is grounding what might otherwise be an abstract conversation by reflecting what is being said on the handout.

Figure 9.5 – Teddy Svoronos displays a PDF and annotates it based on input from his students.

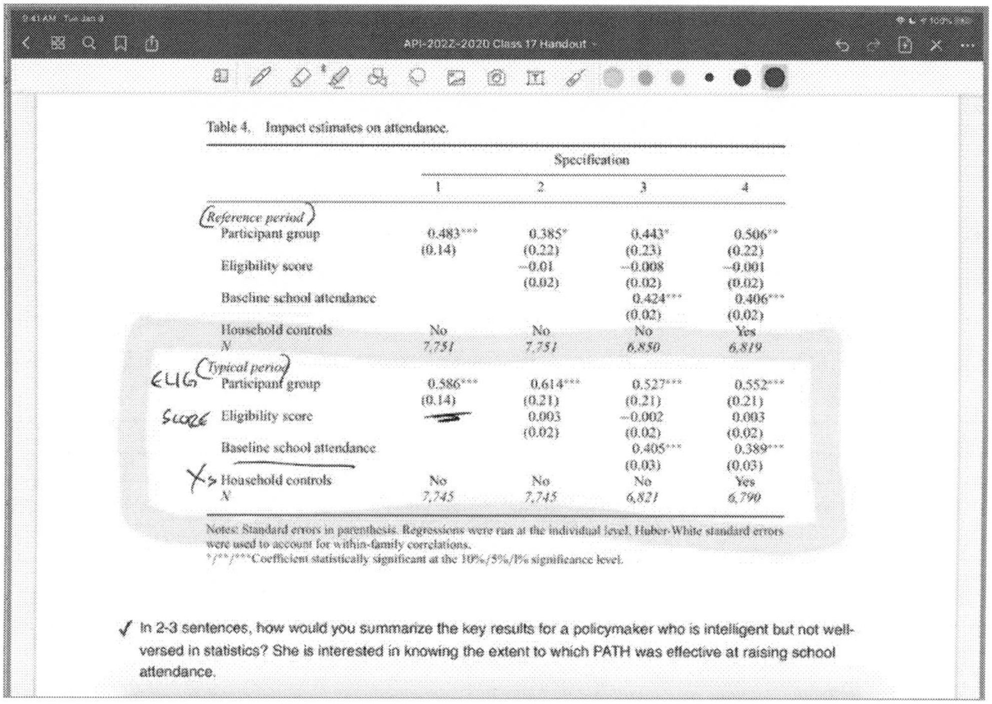

Table 4. Impact estimates on attendance.

	Specification			
	1	2	3	4
Reference period				
Participant group	0.483*** (0.14)	0.385* (0.22)	0.443* (0.23)	0.506** (0.22)
Eligibility score		−0.01 (0.02)	−0.008 (0.02)	−0.001 (0.02)
Baseline school attendance			0.424*** (0.02)	0.406*** (0.02)
Household controls	No	No	No	Yes
N	7,751	7,751	6,850	6,819
Typical period				
Participant group	0.586*** (0.14)	0.614*** (0.21)	0.527*** (0.21)	0.552*** (0.21)
Eligibility score		0.003 (0.02)	−0.002 (0.02)	0.003 (0.02)
Baseline school attendance			0.405*** (0.03)	0.389*** (0.03)
Household controls	No	No	No	Yes
N	7,745	7,745	6,821	6,790

Notes: Standard errors in parenthesis. Regressions were ran at the individual level. Huber-White standard errors were used to account for within-family correlations.
*/**/***Coefficient statistically significant at the 10%/5%/1% significance level.

In 2-3 sentences, how would you summarize the key results for a policymaker who is intelligent but not well-versed in statistics? She is interested in knowing the extent to which PATH was effective at raising school attendance.

Annotating by typing

As indicated before, the two key tools for annotating by typing are:

(1) Zoom's native whiteboard feature
(2) A document on your computer in your preferred software (e.g., PowerPoint, Google Docs, Google Slides, Microsoft OneNote, Microsoft Word, etc.)

(1) Zoom's native whiteboard feature

A simple way of annotating what students are saying is to type into a document. You can do this by using Zoom's native whiteboard feature. This feature is fairly basic at the moment, but it has the advantage that it is integrated into Zoom. So everything happens within Zoom. To use the whiteboard, you simply click on "Share Screen" and select the "Whiteboard" icon.

Figure 9.6 – Activating Zoom's whiteboard feature through "Share Screen"

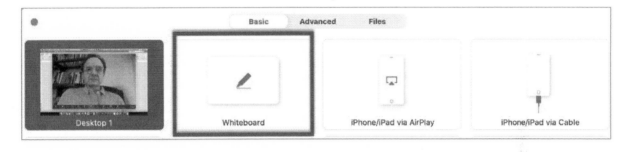

Once you are in the whiteboard, you press "Text" and locate the text box wherever you want on the canvas and start typing. You can also draw and use some of the other options available. If you plan to draw or handwrite using Zoom's whiteboard feature, I recommend connecting a tablet to do this (see "Tech Tips" section below).

Figure 9.7 – Using Zoom's native whiteboard to type text

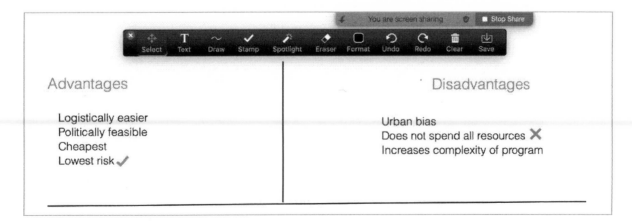

Finally, one advantage of the whiteboard is that your students can also contribute to the output at the same time, though if more than a few students are contributing it can get unwieldy. If you prefer, you can change a setting in Zoom to disable the ability of students to annotate.

(2) Typing on a document in your computer

A more robust option for typing notes during class is to prepare a blank document (e.g., PowerPoint, Keynote, Google Slides, Google Docs, etc.) and show it to students (via "Share Screen") during class while you type and fill it in. If you do this, it might be helpful to prepare templates that set the structure for your notetaking and allows you to more easily type exactly where you want on the page. See "In Practice" below for an example of how to do this with PowerPoint.

In Practice 9.5 – Annotating by typing on PowerPoint

Mitch Weiss teaches entrepreneurship courses at the Harvard Business School (HBS) and uses PowerPoint to annotate what his students say in response to his questions. He has a few slides with text boxes that allow him to quickly decide where he wants to type the comment that the student is making. A key to making this work is to design the note-taking slides in a very intentional way so that he quickly knows where to type the notes. When he is presenting to his students, his PowerPoint presentation is on "Normal" Mode (i.e., not on "Slide Show" mode), so he can type on a slide, and with the sidebar minimized so the slide is as large as possible and so students cannot see later slides. See picture below.

Mitch uses "Share Screen" to share these slides with students at times, but when he wants students to see more of each other (rather than the slides), he continues taking notes in the slides but stops sharing his screen. Throughout the class, he smoothly moves in and out of "Screen Share" mode to allow students to alternate between seeing him and each other ("Screen Share" is off) and seeing the slides he is typing in ("Screen Share" is on).

Figure 9.8 – Mitch Weiss inputs student comments by typing into PowerPoint slides

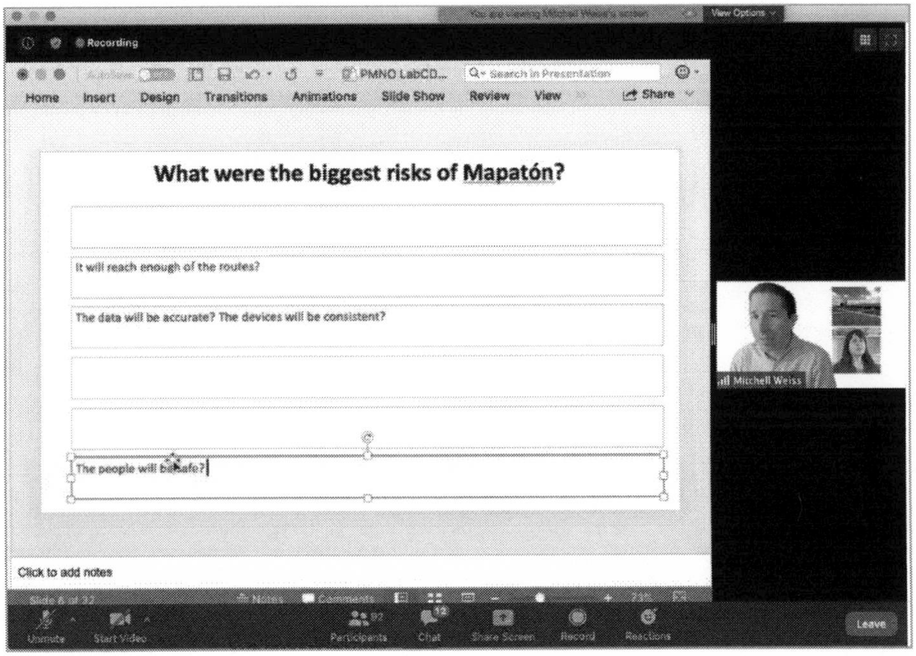

The advantage of using a collaborative solution like Google Slides or Google Docs to record student input is that the students themselves can type the comments directly into the shared document, which makes note-taking a more collaborative enterprise. See chapter 6 – "Work in Groups" for more details on how to use Google Slides in the context of an online live class and chapter 11 – "Building Community" for some ideas on how to use Google Docs to create community.

 **Tech Tips**

- If you plan to annotate using a (physical) flipchart or blackboard in the room where you are teaching, test out your setup with people joining by Zoom who can help guide you to make adjustments, such as how far behind you the flipchart or blackboard should be, which portions are legible, etc.

- If you plan to annotate using a document camera, there are many options in the marketplace. Two key features are resolution and size (it needs to fit on your desk). The <u>companion site</u> has some links of options for you.

- In May 2020, Zoom enabled virtual camera support, which allows you to use a smartphone as a document camera. You need to install software, and there are several options in the marketplace. The <u>companion site</u> has some links with options for you. To share, click on "Share Screen" in Zoom's main toolbar, click on "Advanced" and then click on "Content from 2nd Camera". See image below.

Figure 9.9 – Sharing content from second camera

- If you plan to annotate using an additional device (e.g., iPad, Microsoft Surface, etc.), I suggest connecting that device to your computer by cable and then using Zoom's "Share Screen" feature, rather than using any of the other alternative methods (see "Tech Tips" section in chapter 8 for details).

- If you want to simply use an electronic whiteboard to handwrite student comments, one way to do it is to use Zoom's native whiteboard tool. In theory, you could use the mouse to handwrite but I frankly don't recommend doing it this way as the output rarely looks good and it takes longer to write with a mouse than to write with a pen. If you want to do a fair bit of handwriting and use Zoom's whiteboard for this, I recommend that you consider connecting a tablet (e.g., Wacom, iPad, Microsoft Surface, etc.) with an electronic pen.

Chapter Summary

- Annotating to explain your ideas and/or record student comments can make your classes much more dynamic, engaging, and effective.

- In deciding which tool to use, it is helpful to decide whether you want to annotate by hand or by typing.

- Pre-test the tools to decide what works for you, and test it with friends or colleagues to make adjustments before you use them in the classroom.

PART IV – PUTTING IT ALL TOGETHER

The goal of this final part is to help put some of the things you learned in the previous three parts in a broader context, and to pull things together. Chapter 10 recognizes that online live classes are just a part of the larger ecosystem of online learning, and provides some guidance on how to decide what material should be presented in the live sessions and what should instead be oriented to learning outside of live classes. It then focuses on how to synergize the two types of learning experiences. Chapter 11 is about the crucial goal of building community in your online course. It is organized around practices that you can employ to build community before, during, and outside class. Finally, chapter 12 provides some advice to help you bring together some of what you learned, and offers some ideas for organizing your next steps.

Chapter 10 – Blending Sync and Async

Live online classes, the focus of this book, are just one part of the larger ecosystem of online learning. Just like in an in-person course, your students' engagement with the course does not happen only during the live classes. Students are typically asked to work on a number of things outside class, including readings, homework, online modules, quizzes, projects, reflections, memos, etc. So you are already used to thinking about the balance between work in class and outside of class.

Yet as many of us are transitioning to teach our courses online, we are being encouraged to think more carefully about what material we should try to teach in our live sessions (synchronously) and what material we should ask students to engage with on their own time (asynchronously). This is especially salient now because, as described in chapter 3, many instructors are reporting that they can engage with less material in a live online session than in an equally long live in-person session, which implies that many of us will have to make some hard decisions about what material to keep in our live sessions and what material to ask our students to engage with on their own time or simply cut out. Moreover, some of us are, or will be, in environments where class schedules provide for less live class time than we used to have when teaching students in person.

This chapter starts with a discussion about how to make the decisions of what material should be presented synchronously and what should instead be oriented to asynchronous learning. I then focus on the question of how to leverage students' asynchronous learning to help you conduct better synchronous class sessions.

Key Definitions

- **Synchronous:** You and your students come together in a live session (Zoom).

- **Asynchronous:** Students engage with material on their own time (before and/or after your live session).

Key Decisions

(1) How should I **split** my content into synchronous and asynchronous materials?

(2) How should I **leverage** the asynchronous learning to help me conduct better synchronous sessions?

10.1 How should I *split* my content into synchronous and asynchronous materials?

The answer to this question will depend on what you think is the comparative advantage of a live class session over activities your students can do asynchronously. In other words, what are learning activities where being together makes a difference? This will vary based on your learning goals, the material you teach, and your style of teaching, but below is a table summarizing what I think are reasonably general principles.

Table 10.1 – General Principles of Synchronous and Asynchronous Learning

	Synchronous learning	**Asynchronous learning**
Better when	• Exchange of perspectives among your students is important. • Your students' learning from each other is important. • Dialogue is important. • Your intervention to facilitate or mediate is important.	• You want students to have or to develop a common foundation before your class (especially of basic ideas/concepts). • Knowing your students' perspectives or background on the subject would affect how you conduct your live class.

	• You want to build community.	• The material is such that each student going at their own pace is beneficial for their learning.
		• Students having a good amount of time to ponder and reflect is beneficial.
Typical examples	• A debate • A Socratic discussion • Using polls and asking the students to work in groups to discuss the question in the polls • A discussion of sensitive or emotionally charged issues • An activity that helps build community by being together	• A reading (e.g., book, article, case study, etc.) • A video • A short mini-lecture • Online module where students with little background on the material can go at a slower pace than students with a strong background • A computer programming assignment

The laundry test

At the school where I teach, classes on Zoom generally get recorded so that students can watch them if they cannot attend the live session. I recently asked a student how she decided whether to engage in the live class or watch the recording later. Her answer was revealing. She said, "When I am trying to decide, I ask myself 'Is this a class I could attend while folding my laundry?' If the answer is yes, I watch the recording. If the answer is no, I attend the live session." While I think that, in general, we should design both synchronous and asynchronous experiences that students find so engaging that they cannot fold the laundry at the same time, I think the spirit of this question might help inform your decision of what to reserve for asynchronous learning. If the students can conceivably fold their laundry while engaging in the experience, my advice is to either eliminate it or reserve it for asynchronous learning.

A word of warning

I recently heard an instructor say that he had a lecture that was very boring last year in the classroom and that he was just going to record it so that students could watch it online and he could save some valuable live class time. My thoughts on this, which I shared with him in a much more tactful manner, are: If you think something will be boring in a live session, what makes you think it won't be boring in a recorded session?! More generally, if something will be ineffective on a live session, it will also probably be ineffective in an asynchronous session!

Designing asynchronous material

How to design effective asynchronous material is beyond the scope of this book, but you might find these guidelines helpful:

- **Use the same active learning principles you would use when designing a synchronous experience.** Don't speak to your students for a long time without providing them opportunities to process, internalize, apply, reflect, and make sense of the material. This can sometimes be accomplished by breaking a long recording into several small ones and including questions in between where students can actively engage. Make sure these questions are meaningful and not just busy work.

- **Incorporate into the design of your asynchronous materials opportunities for you to collect information on your students that could inform the design of your live classes.** For example, you could ask a few questions on a survey or quiz that would help you understand which parts of the material your students mastered and which parts they did not. Or you could simply ask students to apply the material to their own lives or to ask open-ended questions about the material. For ideas on what kind of information to collect and how to use it, see the next section.

The companion site has links to some useful resources on how to create asynchronous material, including a "flipping kit" and some concrete examples of asynchronous materials.

10.2 How should I *leverage* asynchronous learning to help me conduct better synchronous sessions?

As indicated above, it is helpful to design asynchronous material in a way that you can collect information about your students that you could then use to conduct better synchronous sessions. The picture below summarizes the three key steps of this "blended learning" approach, where you are blending the synchronous and asynchronous components of your course.

Figure 10.1 – The blended learning process

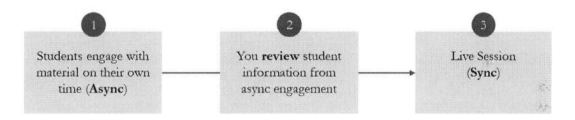

One pattern I have noticed is that some instructors either skip or don't leverage step 2 to its fullest benefit. Below is a table that provides you with some ideas of the type of information you could gather (in step 1) and review (in step 2), and how you might leverage this information in your live classes (step 3). One thing to note is that you often have an asynchronous assignment after the live session (step 3) that could help students solidify their learning and help you adjust the design of the next class, in which case the image below could look like a cycle rather than a sequence.

Table 10.2 – Information you could collect in asynchronous engagement and how you could use it

Information about your students you could collect	How you could use this information
What percentage of them engaged in the asynchronous work	Just because you asked them to do it does not mean they did it. This information could prompt you to remind your students of the work you expect them to do.
What parts of the material they have or have not mastered	This could help you determine how you should allocate your time during class (by spending less time on the material they have already mastered).
What mental models and misconceptions they are bringing to class	Engaging directly with their mental models and misconceptions will help you deepen their learning.
What choices they made on questions where there is no right answer	Knowing how they answered these questions might help you decide how to allocate time in your live session and/or preview which students you want to call on during the live session.
What personal or professional experience they have that is relevant to the material in class	By identifying which students have relevant backgrounds, you could bring their voices to the class at appropriate moments.

How they are applying the material to their lives	You can use the examples they provided to meet students where they are, and help connect the material with things they are interested in.

Two illustrative examples of blended learning in action

These two examples of blended learning in action are meant to spark your imagination, and hopefully help you come up with some ideas for your own live classes.

Example 1 – Leveraging information on what students have mastered and what they haven't

You ask students to engage with some material before class and ask them to respond to a survey or quiz with five multiple-choice questions, each followed by a question asking the students to explain their choice. Below is the data you get from the quiz, where the number at the bottom in each bar represents the percent of students that got that question right and the number at the top the percent that got it wrong. From this, you conclude that students seem to have mastered the material that questions 2, 3, and 4 were testing, but had more difficulty mastering the material corresponding to questions 1 and 5. This would lead you to spend less live class time on content related to questions 2-4, and more class time on content related to questions 1 and 5. Notice that without having collected these data from students in the asynchronous portion of this material, it would have been difficult to make this decision.

Figure 10.2 – Summary of how well the students did in each question in the pre-class quiz

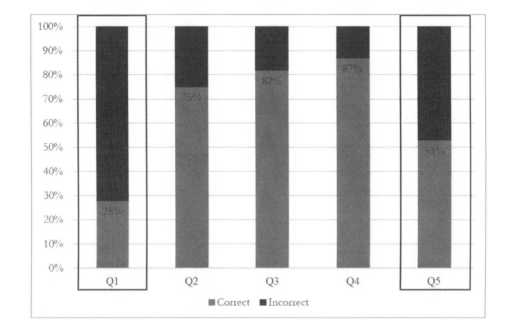

Moreover, you decide to look more closely at question 1. You analyze the justifications provided by those students who answered question 1 incorrectly, and identify some misconceptions that those students have. You also identify some misconceptions after reading the explanations that students who answered these questions correctly gave. Armed with this information, you not only know which parts of the asynchronous materials students are having difficulty with, but also why they are having difficulty. You can now design a class plan that addresses the reasons for these difficulties directly by, for example, giving them a problem in class where the same misconceptions are likely to occur and then helping students see in class (through discussion or other means) why these ways of approaching the problem are not valid. Again, notice that without having collected these data from students in the asynchronous portion of this material, it would have been difficult to take this course of action.

Figure 10.3 – Analysis of justifications that students gave to their answers in one of the multiple-choice questions

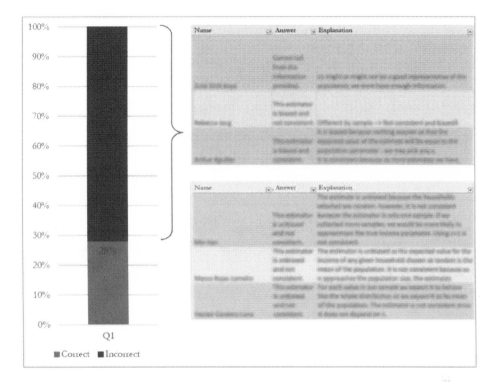

Note: Output on the right was blurred to protect confidentiality of the students.

Example 2 – Leveraging student responses to questions that have no right answer

You ask your students to read an article or a case study, which concludes by asking them to express a preference or take a position, something along the lines of, "Given the evidence presented in this article, which of these three positions do you think is the most reasonable?", where you provide three choices in a multiple-choice question. Or in the case of a case study, the question might be, "If you were in the shoes of the protagonist of the case, what would you decide?" and give students three choices in a multiple-choice question. On either occasion, you would follow the multiple-choice question with a question asking the student to justify the choice they made in a short paragraph.

Table 10.3 – Leveraging student responses to questions with no right answer

Student quiz	Sample data / evidence you might collect
(1) Which option would you choose? ❑ Option A ❑ Option B ❑ Option C	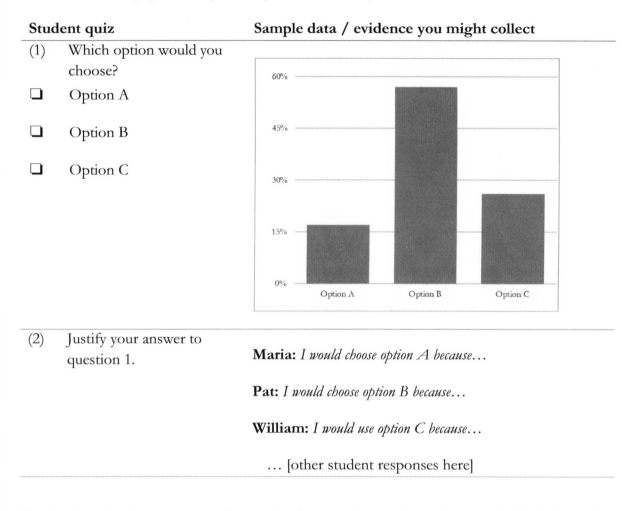
(2) Justify your answer to question 1.	**Maria:** *I would choose option A because…* **Pat:** *I would choose option B because…* **William:** *I would use option C because…* … [other student responses here]

You look at the data corresponding to the first question and see that over half of the students favor option B, and also notice that option A was favored by very few students. You start planning for your live session and think about strategies you might employ by making sure that option A gets a fair representation in the class discussion. You start pouring over the answers to the second question and notice that Ruth, who favored option A, made a very compelling argument for this choice. You decide you will want to call on Ruth when the live session happens. Furthermore, you see that Varun, who voted for option C, has an argument that contrasts with Ruth's argument in an interesting way, and decide that after calling on Ruth you might want to call on Varun to have an interesting debate. Armed with this information, you now go into your live class with a much better sense of where your students are and a good plan to leverage their work in the asynchronous session to conduct a more engaging and vibrant live class session.

When you start the live class session, one challenge you face is how to remember which student favored which option. A low-tech approach is to simply sort student names based on their responses to the multiple-choice question, and print that sheet to draw names from as the discussion unfolds. This is fairly easy to do but will require that you look at the printed sheet of paper while you are teaching your live session.

A higher-tech approach is for you to ask every student to rename themselves in Zoom (see "Tech Tips" section below for instructions on how to do so) according to the option they voted for (e.g., "A – Ruth Levine", "C – Varun Garg", etc.). Now when you look at the participant list, you not only see the names of your students but also how they answered the key multiple-choice question before class. During class, you call on Ruth and Varun as you planned, and a vigorous debate ensues. Then you want to bring someone who favored option B, and look at your participants list in Zoom for a student who voted for option B and notice that three students have their hand up but only Beatriz voted for option B. You call on Beatriz. And so on. Magic happened. The students don't even know how, but you have just orchestrated an engaging discussion in a way that they have rarely seen. And all of this happened because you took some time to think about what information you wanted to know about your student engagement with the asynchronous material, collected this information, and used this information to inform your class plan.

In employing the blended learning approach represented in the illustrative examples, the following questions may emerge:

How do I make sure that students complete what I ask them to do in the asynchronous part?

The answer to this question will depend on your institutional context, but consider making the completion of these "pre-class exercises" part of the course grade. I recommend that you consider grading for completion (or perhaps effort) but not whether they got the right answers.

<u>This sounds great, but I don't have time to do this for every class! What do I do?</u>

You don't have to do this for every class. Like with everything else in teaching, my advice is to start small. Experiment doing this for a couple of classes in your next course. Or to save time, only pose multiple-choice (or numeric) questions without seeking explanatory text. See how it goes. You will learn a lot. You can then tweak and do more the following time. Or you might decide that it doesn't work for you, and not do it again.

<u>When should students do the asynchronous work?</u>

The exact coordination of timing between the synchronous and asynchronous work will depend on many factors, but in general you want the asynchronous work that is meant to inform your class plan to be due sometime before class. If your live class is at 9 am, you could make it due the day before to give yourself some time to go over the information and adjust your class plan. Even if it's due shortly before your class session, you might have enough data by the previous day to at least start making some decisions, and can then make some final tweaks just prior to class.

 **Tech Tips**

- For students to change their name in Zoom, ask them to do the following (see below):

    o   Go to participant list.
    o   Hover over your name.
    o   Click on More.
    o   Click Rename.
    o   Type your new name.
    o   Hit enter.

Figure 10.4 – Instructions for renaming in Zoom by adding the option the student voted for before their name

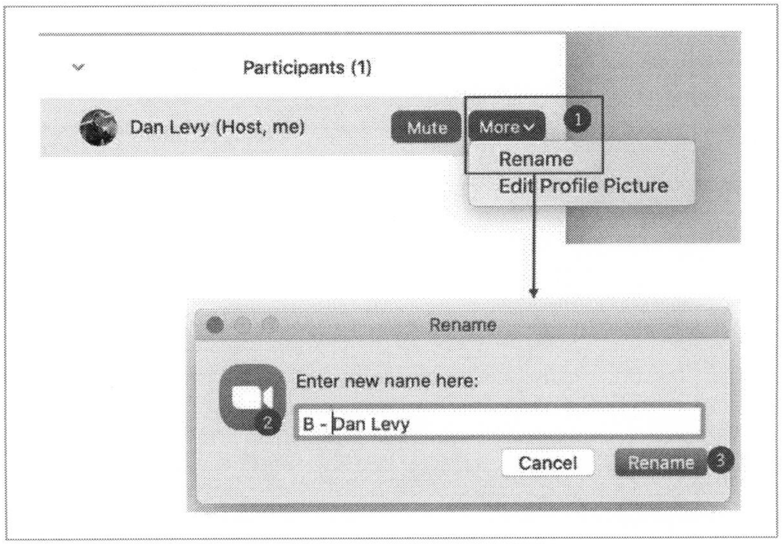

⊙ **Chapter Summary**

- Live online classes are just one part of the larger ecosystem of online learning.

- Key decision #1 – How should I **split** my content into synchronous and asynchronous materials?

  o Remember the laundry test: If the students can conceivably fold their laundry while engaging in the experience, consider eliminating it or reserving it for asynchronous learning.

- Key decision #2 – How should I **leverage** the asynchronous learning in order to conduct better synchronous sessions?

  o Collect data on your students' engagement with the asynchronous material that you can then use to design a better live class.

Chapter 11 – Building Community

A lot has been written about the importance of building community in online learning.[19] Classroom interactions (between your students and you, and among your students) happen almost automatically in the physical classroom, but in the virtual classroom, you have to be much more deliberate about creating these interactions, which are essential to building a strong learning community. The ultimate goal of a strong learning community is to instill a sense of belonging and camaraderie that keeps your students engaged and motivates them to learn and persevere. This chapter focuses on how to build community in online live classes. The table below breaks down community-building into three key goals and provides illustrative examples for ways of reaching each goal. The companion site has some suggested readings and resources for how to build community in online learning more broadly.

Table 11.1 – Community-Building Goals

Community-Building Goal	Examples
Foster opportunities for students to **engage with you**	• Open your virtual classroom a few minutes before class and leave it open for a few minutes after class.
	• Expand your office hours.
	• Explore new forms of office hours (with a group, for example).
	• Conduct more 1:1 correspondence with students.
	• Learn your students' names and basic backgrounds.
	• Collect feedback that informs you how each student is faring as an individual in the course.

Foster opportunities for students to **engage with each other**	• Use ad-hoc small group breakouts. • Design small or low stakes group work. • Design larger or higher stakes group projects. • Provide opportunities for structured peer review/feedback.
Engage students on **co-building the learning community**	• Use icebreakers at the beginning aimed at helping students learn more about each other, especially with regard to experience relevant to the course. • Encourage your students to co-create class participation norms. • Ask students to curate new content and examples for the course.

This table is based on input from my colleague Maria Flanagan, a learning designer who is very thoughtful about community-building and other key aspects of online learning.

The rest of this chapter is organized around practices that you can employ to build community before, during, and outside class.

11.1   Practices to build community before class begins

It is a good idea to invite students to join your live online class a few minutes before class begins. This allows students to see you and each other, engage in some activities that strengthen their sense of belonging and community (see ideas below), test their technology (video, microphone, etc.), and ensure they are ready to learn when the official class time starts.

How long before class starts should you ask students to join? This obviously will vary with your institutional context, personal preferences, and time constraints. But here are two things you might want to keep in mind: First, as indicated above, you have to be much more deliberate about creating interactions online compared to face-to-face teaching, and being available to your students just before (and just after) class is one easy way to do it. Second, one reason you might not be able to be available to your students for much time before class when you teach in a physical classroom is because someone else is occupying the classroom before you. But in a virtual classroom, this is not a constraint! You and your students of course would still need to be available before class starts, but if you are, this seems like a great way to engage with them, build community, and maximize the chances of a productive class.

What do I do before class begins?

You can do a number of things to build community before class begins, including asking ice-breaking questions, playing music, and getting participants to express themselves in various ways. Below are examples for each of these activities. Whatever you decide to do, I suggest that your goals be simply to be present, connect with your students, and create opportunities for them to connect with each other.

 In Practice 11.1 – Using ice breaking questions before class begins

Carrie Conaway is a senior lecturer at the Harvard Graduate School of Education, and was the chief strategy and research officer for the Massachusetts Department of Elementary and Secondary Education (DESE). She opens her live session a few minutes before class is scheduled to begin, warmly welcomes the students as they join, and asks them an ice-breaking question (e.g., which Game of Thrones character would you like to be and why?) and encourages them to respond by voice or chat.

In Practice 11.2 – Using music before class begins

Teddy Svoronos, a lecturer at the Harvard Kennedy School by day and a jazz musician by night, loves music. He shares his love for music with his students by playing songs from a Spotify playlist that he created for his class. He invites students to contribute to this playlist, which makes this one of the things that his students create as a community in his class. He also encourages students to write something in the chat before class begins. See his first slide below, which he displays to students before class. To learn how to share music with your students during a live session, see "Tech Tips" section below.

Figure 11.1 – Sample slide inviting students to chat before class begins

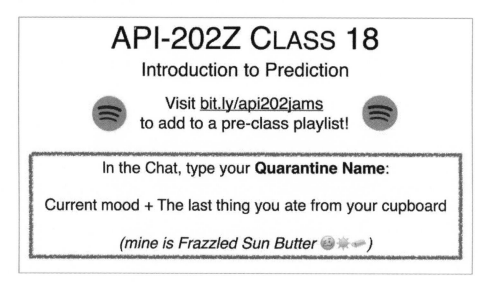

11.2   Practices to build community during class

Some of the practices used to create community during a class in a physical classroom can be used in a virtual classroom as well. For example, getting to know your students and connecting with what they bring to the class can help them learn more deeply and create community. Providing opportunities for students to interact with each other in breakout rooms (see chapter 6) is another effective way of advancing both learning and community-building. Having a ritual at the end of class (such as everyone unmuting themselves and saying something, or everyone writing something on the chat) can also help create community. And after class officially ends, consider staying for extra time to engage students in informal conversation and/or use informal breakout rooms to provide your students with easy opportunities to socialize with each other. The "In Practice" sections below offer some other ideas for building community online during your class.

In Practice 11.3 – Asking students to rename themselves and using breakout rooms

Anna Shanley, Maddie Meister, and James Brockman work in Executive Education at the Harvard Kennedy School, where participants come from all over the world. In the online programs they are involved in, they ask participants before classes start to rename themselves in Zoom (see "Tech Tips" section in chapter 10 for instructions on how to do this) using the following convention: "First Name & Last Name; Location; Organization; Job Title" (e.g., "Anna Shanley; Cambridge MA; HKS Program Director"). This allows participants and instructors to see information about each other on the screen, which helps people get to know one another more quickly than they would otherwise. This practice also enables the instructor to use this information before or during their teaching (e.g., "Mary, I heard there is a heat wave in Arizona; how are you doing?" or "Sarah, given your background in finance, what would you do if you were presented with this financial proposal?"). For instructions on how to rename on Zoom, please see "Tech Tips" in chapter 10.

They also use breakout rooms very effectively in their programs. Some of the breakout rooms are random (which allow participants to network with each other) and others are deliberate (which allows participants to get to know themselves better and produce work that they can share with the class). When they use random breakout rooms, they typically instruct participants to introduce themselves once they are in the breakout room and answer an ice-breaking question. See example slide below. To give you a sense of what they have been able to accomplish, one participant at the end of one of their programs said, "I have attended one course on campus, and I actually found that I bonded with more of my classmates in this setting than in a face-to-face setting; the breakout room discussions were very valuable, as I was able to connect with people with whom I would not otherwise have known I had to so much in common."

Figure 11.2 – Sample slide giving students instructions before sending them to breakout rooms

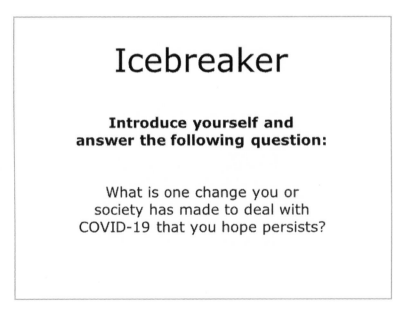

In Practice 11.4 – Using check-in questions about a common experience at the beginning of class

Zoe Marks, a Lecturer at the Harvard Kennedy School, begins every session with a poem on the screen (not read aloud), and with a check-in question. In her words: "Some days I asked what they wanted to see come out of the pandemic crisis for the better, some days I asked what they were sad or angry about, some days what made them hopeful, and sometimes anyone they wanted to keep in their/our thoughts. It provided an icebreaker that touched on our common experience, while hopefully making room for people's specific circumstances and need to either be seen or be quiet but in community with others experiencing similar good and bad challenges."

In Practice 11.5 – Using interactive verbal roll to create community

Kathy Pham teaches a course on Product Management and Society at the Harvard Kennedy School, and uses a variety of tools to create community in her class. At the beginning of every class, she asks students to write their name, pronouns, location, and an answer to one question for that class in a Google Doc. The question can be about the material for that class (see example below) or a more social question (e.g., favorite band name, favorite childhood food, etc.). If it is about the material for that class, she might call on some of the students who provided answers during class. She sometimes begins class doing a verbal roll call where she says their names, and they respond with what they wrote in the Google Doc. She said about this practice, "I did this because I wanted everyone to feel seen and it felt nice to hear people's names and hear them respond." Beatriz Vasconcellos, one of the students in Kathy's class, commented that this sort of interactive verbal roll call "helped make students engage in class right from the beginning and build community." See example below of her roll call for one of her classes.

163

Figure 11.3 – Kathy Pham uses a Google Doc to create an interactive roll call

---

Wednesday, April 15th, 2020

Add yourself to the Google Doc:

(I will call on some of you to discuss the product partnership below)

Name | pronouns | current location | Name an interesting product partnership (two products, teams, companies, etc., that joined together to make a product better)

Evan Collins | they/them | Somerville, MA | Starbucks + Spotify

Scarlet Roman | she/her/hers | Allston, MA | Most stores + Apple Pay

Kimora Christensen | they/them | Los Angeles, CA | CVS and Target

Karla Hamilton | she/her | Cambridge, MA | Pokemon + Niantic

…

Hannah Chang | she/her | Taipei, Taiwan | EVA Airlines + Hello Kitty

Nicodemo Blanco | he/him | Florida | Apple + Google COVID

Katelyn Hanson | she/her | Lakewood, CO | Cover Girl + Lucasfilm

Claudia Cochran | Cambridge, MA | Samsung and Google

Chloe Frost | she/ her | Cambridge, MA | Spotify and Hulu

Pascual Carrasco | Cambridge, MA | Lyft + Capital One

Note: Student names have been changed to protect confidentiality.

In Practice 11.6 – Building community during class one small gesture at a time

Allison Shapira, who teaches communications courses at the Harvard Kennedy School, does a lot of small things to build community in her classroom. Here is how she frames her efforts: "In my research preparing to teach virtually at Harvard, I found that one of the biggest reasons students drop out of online courses is due to a feeling of loneliness and isolation, so I begin every class by asking students to say hello in the chat and say how the weather is (or the mood is) in their country or city. Then, I might cold call someone to take themselves off mute to answer. I also learn how to say hello to students in their own languages which always brings a smile to their face. To keep students engaged, I'll stand up while teaching and keep the camera lens at eye level so that I can both create and transmit energy through the camera lens to my students. When things go wrong, I laugh at the situation and show students you can laugh when things go wrong. I use "Gallery View" so I can see students' reactions; if I notice one of my students' children wandering in the frame of view, I might stop and say, "Steve, is that your baby daughter? She is beautiful!" It makes the student feel seen and the other students feel that I'm paying attention to them."

11.3   Practices to build community outside of class

A lot of community-building happens outside class without your intervention. Students might create WhatsApp or Slack groups to interact with each other. They might organize parties, dinners, outings, and other events (virtually or in person). And your school/institution might organize events for your students, such as orientation sessions, guest speakers, workshops, and so on. These community-building efforts don't negate the need for you to think about building community outside class, but knowing about them might inform where you focus your energies. For example, devoting part of a class session to having your students introduce themselves to each other does not make sense if they have already done so at an event organized by your institution. Below is one example of the types of activities that can be done outside class to build community.

 In Practice 11.7 – Building community outside class

Dana Born, Horace Ling, Keisha Mayers, Susan Tuers, and Cindy Wong recently led a 4-week long online executive education program at the Harvard Kennedy School. The program is intense, and they creatively used a number of activities to help build community. "One of the biggest challenges of the online space for education is the difficulty in creating a community and network. When we talk to prospective participants, the most common concern is that they won't be able to connect with the faculty and their classmates to the level that they would during an in-person program," Horace reported.

They were successful at building community by implementing multiple activities, each of which attracted different participants, including:

• Online 5K Run – Participants went for a run, pausing from time to time to share the views of where they were. Many who didn't run also enjoyed and watched the exercise. Participants were in Washington DC, Seattle Washington, Ottawa Canada, Lexington Massachusetts, and Provincetown Massachusetts.

- Online Talent Show – Participants volunteered to sing, do yoga, tell jokes, or share other talents.

- Mentoring Online – Each participant was assigned one co-mentor to meet with.

- Live Cooking Show/Networking with Alumni – Participants cooked along with the Harvard Director of Culinary Operations. Afterwards, they ate lunch with each other and with the alumni that were invited to join.

- Affinity Groups (Organized by Participants) – Various topics that participants wanted to talk about that others were interested in as well.

- Lunch Chats with Faculty – Faculty met with participants for an informal hour during lunch.

The list above might not be directly applicable to you, but hopefully it helps spark some ideas for activities that you can use to build community in your course.

This chapter ends with an "In Practice" section describing the approach and many practices to build community adopted by my colleague Marshall Ganz and his teaching team. While it might be difficult for many of us to achieve the exceptional sense of learning community that Marshall has achieved in his online classes, I think we can all learn from his practices and adopt a few for ourselves.

In Practice 11.8 – Building community online: an example of mastery

Marshall Ganz teaches courses in leadership, organizing, and public narrative at the Harvard Kennedy School. He has successfully taught thousands of students online, and to me represents the ultimate example of what can be accomplished online when the instructor is as intentional and deliberate as Marshall is. One of the practices he is most distinguished at is building community. He argues, "You can do online whatever you can do in a room as long as you establish the relationships." To give you a sense of how successful he is at building community, on the last day of his online learning programs (which typically meet twice a week for several weeks), students do brief presentations about their learning in the groups they have been working with throughout the program. There is laughter, joy, tears, and emotional connection like I have rarely seen in a classroom. Remember these are online courses, and most of these students have never met face to face!

Marshall often tells the story of a student from Norway who took one of his online courses on public narrative. Sometime later, she visited the Kennedy School and asked Marshall for permission to sit in one of his in-person classes on the same topic. At the end of class, Marshall asked her what she thought about the class. She responded, "I think online was more intimate." Marshall was surprised by this answer and probed. She said, "Well, I am sitting in a classroom looking at the backs of everyone's heads. But online I see everyone's faces, and I can see tears, and I can see laughter, and I realize what I am a part of." In reflecting about this encounter, Marshall says, "Our faces communicate so much of our emotional language. It really struck me; it's not something I would have thought of before. But it's the truth."

Describing all that Marshall does to achieve this sense of community in his courses is beyond the scope of this book, but these are some of the practices I noted as crucial:

- Establishes strong community norms. See example slide below.

- Acknowledges from the beginning that "Nothing will be perfect; we are all here to learn. We are partners in learning," and follows through on this by asking students for quick feedback after every class session.

- Students work in fixed groups for some activities and random groups for others. Fixed groups help build community; random ones help build networks.

- Groups are asked to set their own norms, they come up with a group name and a group chant. This helps group members be accountable to each other.

- Each member of the teaching team is responsible for the learning of their own section of about twenty students. They meet with each student one on one, facilitate discussion, coach students, and evaluate their work. They work with the same section for the entire course, guide learning, sustain motivation, and enable peer accountability.

- Conducts purposeful and highly structured facilitation where all students are encouraged to participate in a wide range of forms.

- Ends class by asking everyone to unmute themselves and applaud.

Marshall is extremely generous with sharing the lessons of what he has done. The companion site has links to several of his resources, including a video presentation he gave on his online teaching practices.

Figure 11.4 – Marshall Ganz's slide summarizing use of community norms

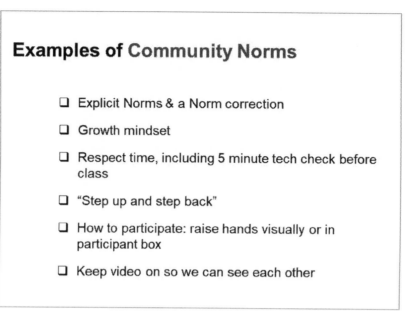

Examples of Community Norms

❑ Explicit Norms & a Norm correction

❑ Growth mindset

❑ Respect time, including 5 minute tech check before class

❑ "Step up and step back"

❑ How to participate: raise hands visually or in participant box

❑ Keep video on so we can see each other

 **Tech tips**

- To play music for your students in Zoom, you have two options: the first one involves simply playing music without sharing anything else, and the second one involves sharing music while you are sharing something else (e.g., a slide).

- Option 1: Just share the music.

  o Click on "Share Screen" in Zoom's main toolbar.
  o Click on "Advanced".
  o Click on "Music or Computer Sound Only".
  o Click on "Share".

Figure 11.5 – Sharing music for your students in Zoom (and nothing else)

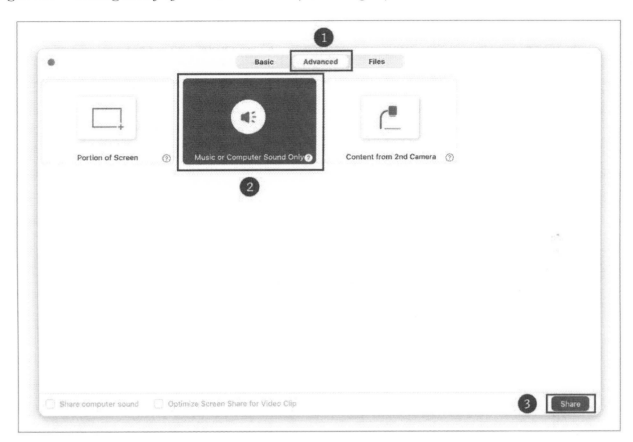

- Option 2: Share music and something else (e.g., a slide).

    o Click on "Share Screen" in Zoom's main toolbar.
    o Click on the other thing you want to share (e.g., Powerpoint, Keynote, etc.).
    o Check "Share Computer Sound" at the bottom.
    o Click on "Share".

Figure 11.6 – Sharing music along with something else with your students in Zoom

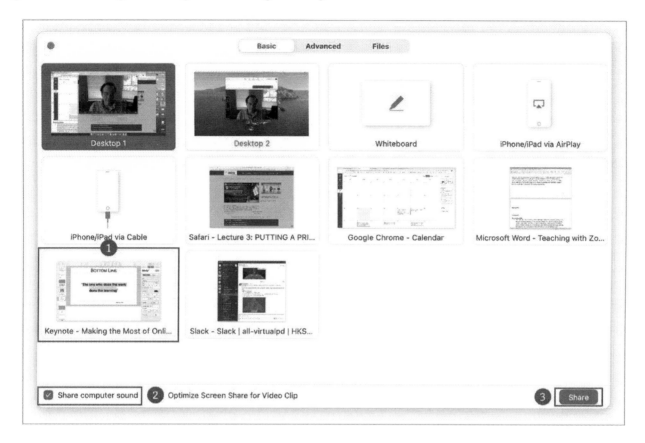

📖 Chapter Summary

- Building community is crucial for a successful online learning experience.

- You have to be much more deliberate about building community online than in person.

- Building community involves:

  - Fostering opportunities for students to engage with you
  - Fostering opportunities for students to engage with each other
  - Engaging students on co-building the learning community

- Think about what you can do to build community:

  - Before class
  - During class
  - Outside class

- The most important aspect of building community is to foster human connection. The particular activities you employ are less important than your desire to do this genuinely.

Chapter 12 – Next Steps

This final chapter provides some advice to help you bring together some of what you learned, and offers you some ideas for organizing your next steps. Below are some of the key lessons I hope you took from the book. Before you review them, I suggest that you take 5 minutes right now to write down the key ideas that *you* took from the book. If you kept notes while reading, this might be a chance to add any ideas you have recently had. Doing this will help you learn them more deeply than you would if you immediately read the list below, and it will increase the likelihood that you implement your ideas.

Part I: Key ideas – Summary

- Transitioning from a physical classroom to a virtual one represents a loss. Acknowledging this loss is important.

- Teaching online, just like teaching in physical classrooms, is all about the learning of your students. The technology, including Zoom, is just the vehicle to reach your goals.

- Think about what pedagogic principles underlie your teaching practices, and how best to apply these principles in designing and teaching your live online classes.

- The key pedagogic principles underlying the practices and advice advanced in this book are:

  (1) Be student-centered.
  (2) Plan for active learning.
  (3) Begin with the end in mind.
  (4) Use online teaching to its comparative advantages.
  (5) Teachers are made, not born.

Part II: Ways your students can engage – Summary

- There are more ways of engaging your students in an online classroom than in a physical one. Leverage this fact to help you reach your goals.

- Key ways your students can engage (with key Zoom tool in parenthesis):

  o Speak (Raise hands)
  o Vote (Polling)
  o Write (Chat)
  o Work in groups (Breakout Rooms)
  o Share their work (Share Screen)

- Conversations tend to take longer in a virtual classroom than in a physical classroom. Take this into account when you plan your class.

- Polls can be a very helpful tool to engage students, assess where they are, and teach in a more flexible manner.

- Chat is a very quick and efficient way to find out what is on your students' minds, but if you decide to use it, you will benefit from establishing and communicating norms around its use.

- There are a lot of benefits to having students work in small groups during part of your online live classes. For this to work well, being explicit about the task you want students to accomplish (e.g., answer a question, produce a deliverable, etc.) and the time they have available is crucial.

- Whenever you ask students to produce work, think about the potential benefits of having some of them present this work in a live class session and the ways in which this can be done most effectively.

Part III: Ways you can engage – Summary

- You can use Zoom to present slides, videos, websites via your internet browser, and pretty much anything on your computer.

- Remember that presenting slides for a long time without having students engage in their learning is unlikely to be effective. Plan for interaction every few minutes of class.

- Be aware that the moment you share your slides, they are likely to take over a large share of your students' screen and attention. Do so judiciously and stop sharing your slides in segments of the class where you don't need them.

- When presenting, try to keep your eyes on your students rather than on your slides. If you have access to a second monitor, this goal becomes easier to achieve.

- Annotating to explain your ideas and/or record student comments can make your classes much more dynamic, engaging, and effective.

- In deciding which tool to use, it is helpful to decide whether you want to annotate by hand or by typing.

- Pre-test the tools to decide what works for you, and test it with friends or colleagues to make adjustments before you use them in the classroom.

Part IV: Putting it all together – Summary

- Live online classes are just one part of the larger ecosystem of online learning.

- Think about your students' work falling into two categories:

- Synchronous: You and your students come together in a live session (Zoom).

- Asynchronous: Students engage with material on their own time (before and/or after your live session).

- Two key decisions in the larger online learning ecosystem:

  o Key decision #1 – How should you split your content into synchronous and asynchronous materials?
  o Key decision #2 – How should you leverage the asynchronous learning to help you conduct better synchronous sessions?

- Building community is important in general, and even more important in online teaching.

- Building community online requires a more deliberate effort than in person.

- Think about what you can do to build community before, during, and outside class

12.1   Next steps

The next steps in your journey depend partly on where you are:

1. Before starting your online course
2. During your online course
3. After your online course

Below are some thoughts for each of these stages.

1 - Before starting your online course

My three key recommendations for getting ready for an upcoming online course are:

- Get some practice.
    - Practice alone with Zoom.
    - Do one or more dry runs of your first class with a group of friends or colleagues and get feedback. This will help you identify a number of opportunities for improvement. See sample list below.
    - Tweak your practices and setup.

Table 12.1 – Sample list of issues you might discover in your dry run and possible ways of addressing them

Feedback received	Way of addressing it
We cannot see your face well.	• Play with the lighting in your room (e.g., add a lamp). • Consider getting a better camera or using your phone for video.
You are not making eye contact with us.	• Tweak your desk setup to make it easier to look at the camera. • Change the way you arrange your Zoom windows to make it easier to look into the camera. • Change the placement of your camera (if portable).

You did not notice that I had my hand up for a few minutes.	• Make sure the participants list is always visible to you; this is easier if you have a second monitor. • Train yourself to keep an eye on the participants list.
Instructions for our breakout room were unclear.	• Investigate what was unclear. • Create a slide with instructions that addresses what was unclear.
You looked frazzled at the beginning	• Consider developing a routine where you log in to your online classes before of everyone else, test your equipment, try out the things you plan to do, so that when your students arrive you are ready to engage with them. • If you can, try to get someone to help you with some aspect of the technology (e.g., monitor the chat)
Our breakout room ended very suddenly.	• Tweak the Zoom settings so that people get a one-minute warning before the breakout rooms end.
Your energy level was low.	• Consider changing your setup to increase the chances of projecting your energy. In general, standing is preferable to sitting.
We couldn't see what you were writing on your blackboard.	• Consider moving the board to be closer to the camera. • Consider getting better chalk/markers. • Consider using some other technology for annotating (Zoom's native whiteboard, a tablet, etc.).

- Establish norms for your students.

  o Develop norms for how students will engage in your online class. See table and sample slides below. Slides are also available on the companion site.

  o If you plan to use chat, make sure you establish some norms around its use.

  o Think about what you will do if/when norms are broken the first time.

Table 12.2 – Zoom-related norms

Aspect	My norms*
How to speak in class	<ul><li>Use "Raise hand to indicate you want to speak.</li><li>Use "Lower hand" after you have spoken or if you no longer want to speak.</li><li>Unmute yourself before speaking, and then mute yourself.</li></ul>
Use of chat	<ul><li>Use privately to a designated person for technical support.</li><li>Use publicly only for comments/questions directly related to the discussion in class at that moment.</li><li>I will pause at various points in class to check it.</li></ul>
Use of audio	<ul><li>Please use "Mute" as your default.</li><li>Please "Unmute" yourself when you want to speak.</li><li>If possible, use a headset to improve audio quality.</li></ul>

Use of video feed	• Unless you have a good reason not to, please keep your video on so you can be seen by your peers and me.
Recording sessions	• I will record sessions for you and me, but these are not to be shared with anyone else.
	• If you cannot come to a live session, watch the recording and answer the discussion questions.
Start time	• I will open the "live classroom" 10 minutes before official start time.
End time	• We will end class on time.
	• I will stay in the Zoom meeting for about 15 extra minutes for anyone who wishes to have an informal follow-up discussion or simply wants to chat for a little bit. This is entirely optional.

*These are not meant to represent what the right norms are for you, but rather to give you a concrete sense of the kinds of things that you want to be explicit about when establishing and communicating your Zoom-related norms.

Figure 12.1 – Sample slide for norms to begin your live session

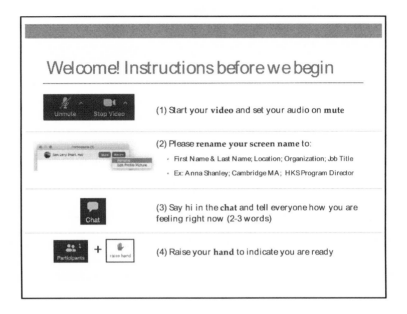

Figure 12.2 – Sample slide for norms related to use of Zoom features

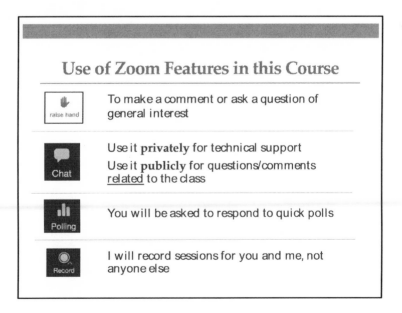

- If possible, get help.

  o There is a lot to keep track of when you are teaching with Zoom. If you can have someone (ideally a teaching assistant, but it could be a colleague or even a student) help you during your live sessions, it can make the whole process more manageable.

  o A few possible roles for your helper:
    - Assist students if they are having technical issues

    - Run breakout rooms

    - Monitor chat

    - Alert you to issues you are not aware of (you are on mute, the slides are not projecting, etc.)

    - Launch polls

2 - During your online course

- Have a colleague observe you teach and ask for feedback. See sample feedback below to give you a sense of the kinds of things you can learn both as an observer and as the person being observed.

- Observe a colleague teach. Learn from them and give them feedback if they want it.

- Collect feedback from your students as to how things are going. You can do this with an anonymous poll or survey.

- Collect evidence about what your students are learning.

- Make adjustments based on the feedback and evidence you collect.

- Assess whether your Zoom-related and other norms need tweaking.

- Write some reflections after every class regarding what went well and what could be improved. This will help you teach much better next time!

Subject: Feedback

Dear Jane:

Thank you very much for the opportunity to let me observe you teach and learn from you. As requested, below is some feedback that I hope can be helpful to you. Thanks again and best regards,

Dan

======

Strengths

- Highly organized class where time was used very productively. Just as a small sample: Telling students on the chat "Class will start in two minutes" is a great way of making sure everyone is ready to start.

- You warmly welcomed your students.

- Great use of roadmaps. Constantly telling students where you were going and marking the transitions very clearly. This seems particularly important online.

- You have a conversation with each student who participates. You seek to understand, probe, investigate, which I think leads students to prepare better for what they plan to say.

- Use of yes/no quick polls, and being able to call on a student according to their response to the poll.

- I liked the idea of the icebreaker for the breakout rooms. I can imagine this being very effective in courses where students don't know each other.

- You know your students and it shows. Welcoming a student from India, telling Jim to bring his higher ed experience, telling Ankur "you are in NY."

- Great at connecting student comments.

- You care about your students and it shows.

Observations/Suggestions/Questions

- Hard to get students to participate in the large group discussion. And you tried!

- I noticed that after the polls you tended to call on someone who had the right answer. This seems great in terms of efficiency and probably helps people feel less intimidated with the cold call, but provides you with fewer opportunities to learn where students are confused and what the key misunderstandings are. I wonder if you have thought about calling students who have the wrong answer before calling on students with the right answer.

- Why did most people not have their video on?

- You asked several questions that forced the students to make a decision between two choices. I wonder if you could transform some of these into quick polls where people can answer Yes/No, and you can interrogate students who chose each of the options.

- I wonder if you could use visuals more. Table 1 exhibit for example.

3 - After your online course

- Reflect on what went well and what could be improved.

- Develop a plan for next time.

- Observe colleagues teach. You will learn a lot.

- Think about adjustments to your Zoom-related norms that might be warranted next time.

- Think about adjustments to your technology and workspace that would make a big difference in your ability to teach and that fit your budget.

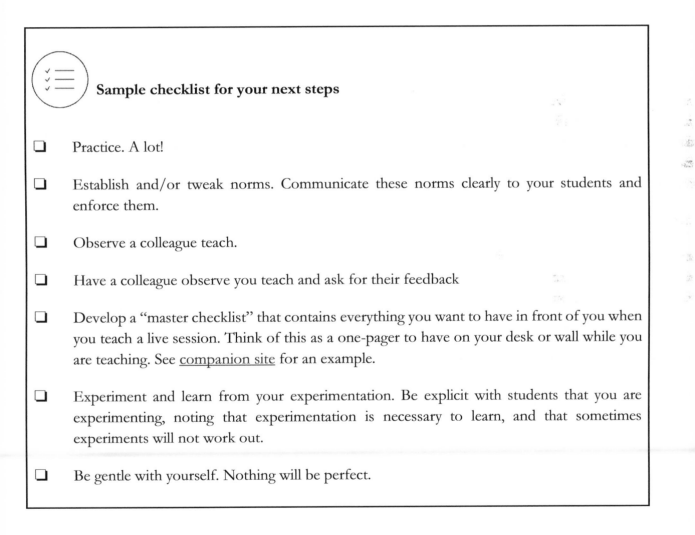

Sample checklist for your next steps

❏ Practice. A lot!

❏ Establish and/or tweak norms. Communicate these norms clearly to your students and enforce them.

❏ Observe a colleague teach.

❏ Have a colleague observe you teach and ask for their feedback

❏ Develop a "master checklist" that contains everything you want to have in front of you when you teach a live session. Think of this as a one-pager to have on your desk or wall while you are teaching. See companion site for an example.

❏ Experiment and learn from your experimentation. Be explicit with students that you are experimenting, noting that experimentation is necessary to learn, and that sometimes experiments will not work out.

❏ Be gentle with yourself. Nothing will be perfect.

12.2   In closing

I hope that reading this book enabled you to develop some ideas to help you improve how you teach online live classes. This was certainly my goal. If you come up with some ideas that you think would be of interest to others, please share them with me on the companion site, as I would love to help you share them with others (with full credit to you, of course). As you put these ideas into practice, some of them are likely to work well and some are likely to fail. This is always bound to happen when we experiment with new technologies and teaching approaches. I think this is OK as long as we can learn from our experimentation. And let us remember that technology is just a vehicle to help us reach our learning goals. As educators, our mission is to help our students learn, grow, and develop. What a noble profession we are in. Thank you very much for your time.

About the Author

Dan Levy has been a faculty member at Harvard University for over 15 years, where he has held various positions related to promoting excellence in teaching and learning. He currently serves as the faculty director of the Public Leadership Credential, the Harvard Kennedy School's flagship online learning initiative. He co-founded Teachly, a web application aimed at helping faculty members to teach more effectively and more inclusively. He has won several teaching awards, including the university-wide David Pickard Award for Teaching and Mentoring. He is passionate about effective teaching and learning, and enjoys sharing his experience and enthusiasm with others.

Notes To This Book

[1] General references for teaching online
- Bates, A. T. (2018). Teaching in a digital age: Guidelines for designing teaching and learning.

- Boettcher, J. V., & Conrad, R. M. (2016). The online teaching survival guide: Simple and practical pedagogical tips. John Wiley & Sons.

- Caulfield, J. (2012). How to design and teach a hybrid course: Achieving student-centered learning through blended classroom, online and experiential activities. Stylus Publishing, LLC..

- Darby, F., & Lang, J. M. (2019). Small teaching online: Applying learning science in online classes. John Wiley & Sons.

- Nilson, L. B., & Goodson, L. A. (2017). Online teaching at its best: Merging instructional design with teaching and learning research. John Wiley & Sons.

[2] Bruff, D. (2019). Intentional Tech: Principles to Guide the Use of Educational Technology in College Teaching. West Virginia University Press.

[3] "The coronavirus allows us to reimagine the college experience" By Bharat Anand The Boston Globe, Updated June 8, 2020, 3:01 a.m.

[4] References on active learning
- Bransford, J. D., Brown, A. L., & Cocking, R. R. (2000). How people learn (Vol. 11). Washington, DC: National academy press.

- Deslauriers, L., McCarty, L. S., Miller, K., Callaghan, K., & Kestin, G. (2019). Measuring actual learning versus feeling of learning in response to being actively engaged in the classroom. Proceedings of the National Academy of Sciences, 116(39), 19251-19257.

- Freeman, S., Eddy, S. L., McDonough, M., Smith, M. K., Okoroafor, N., Jordt, H., & Wenderoth, M. P. (2014). Active learning increases student performance in science, engineering, and mathematics. Proceedings of the National Academy of Sciences, 111(23), 8410-8415.

- Michael, J. (2006). Where's the evidence that active learning works? Advances in physiology education.

- Prince, M. (2004). Does active learning work? A review of the research. Journal of engineering education, 93(3), 223-231.

[5] Doyle, Terry (2008). Helping students learn in a learner centered environment: A guide to teaching in higher education. Sterling, VA: Stylus.

[6] *A Nobel Laureate's Education Plea: Revolutionize Teaching*. Westervelt, Erich, April 14, 2016.

[7] Wiggins, Grant, and McTighe, Jay. (1998). *Understanding by Design*. ASCD.

[8] "The coronavirus allows us to reimagine the college experience" By Bharat Anand The Boston Globe, Updated June 8, 2020, 3:01 a.m.

[9] *"What's that again? The intrinsic psychology of "Zoom fatigue"*," The Economist, May 16, 2020.

[10] Benefits of wait time
- Cain, S. (2013). Quiet: The power of introverts in a world that can't stop talking. Broadway Books.

- Dolan, E; Collins, J. (2017). *We must teach more effectively: here are four ways to get started*. Molecular Biology of the Cell, 26(12)

- Reda, M. M. (2010). *What's the Problem With Quiet Students? Anyone? Anyone? The Chronicle of Higher Education*. https://www.chronicle.com/article/Whats-the-Problem-With-Quiet/124258

- Rowe, M. (2003). Wait-Time and Rewards as Instructional Variables, Their Influence on Language, Logic, and Fate Control: Part One – Wait Time. Journal of Research in Science Teaching.

- Ruhl, K. L., Hughes, C. A., & Schloss, P. J. (1987). *Using the pause procedure to enhance lecture recall.* Teacher education and special education, 10(1), 14-18.

- Schwegman, J. (2013). *Engaging introverts in class discussion–Part 2.* Stanford Teaching Commons, https://teachingcommons.stanford.edu/teaching-talk/engaging-introverts-class-discussion-part-2

- Tobin, K. (1980). *The Effect of an Extended Teacher Wait-Time on Science Achievement.* Journal of Research in Science Teaching, 17(5)

[11] Benefits of polling

- Davis, B. G. (2009). *Tools for teaching.* John Wiley & Sons.

- McKeachie, W., & Svinicki, M. (2013). *McKeachie's teaching tips.* Cengage Learning.

- Prince, M. (2004). *Does Active Learning Work?* A Review of the Research. Journal of Engineering Education, 93(3), 223-231.

- Crouch, C. H., Watkins, J., Fagen, A. P., & Mazur, E. (2007). *Peer instruction: Engaging students one-on-one, all at once.* Research-based reform of university physics, 1(1), 40-95.E.

- Mazur, E. (1997). Peer *instruction: A user's manual.* Prentice Hall series in educational innovation.

[12] Levy, Yardley and Zeckhauser, *"Getting an Honest Answer: Clickers in the Classroom"* Journal of the Scholarship of Teaching and Learning, Vol 17, No 4, October 2017.

[13] References suggesting that most people are not good at multitasking

- Ophir, E., Nass, C., & Wagner, A. D. (2009). *Cognitive control in media multitaskers.* Proceedings of the National Academy of Sciences, 106(37), 15583-15587.

- Sana, F., Weston, T., & Cepeda, N. J. (2013). *Laptop multitasking hinders classroom learning for both users and nearby peers.* Computers & Education, 62, 24-31. https://www.sciencedirect.com/science/article/pii/S0360131512002254

- Sanbonmatsu, D. M., Strayer, D. L., Medeiros-Ward, N., & Watson, J. M. (2013). Who multi-tasks and why? Multi-tasking ability, perceived multi-tasking ability, impulsivity, and sensation seeking. PloS one, 8(1), e54402.

- Ehrlinger, J., Johnson, K., Banner, M., Dunning, D., & Kruger, J. (2008). *Why the unskilled are unaware: Further explorations of (absent) self-insight among the incompetent.* Organizational behavior and human decision processes, 105(1), 98-121.

[14] *"If You Think You're Good At Multitasking, You Probably Aren't"* January 24, 2013. NPR. https://www.npr.org/sections/health-shots/2013/01/24/170160105/if-you-think-youre-good-at-multitasking-you-probably-arent

[15] Positive effects of working in groups
- Johnson, D.W., Johnson, R. T, & Smith, K.A. (1991a). *Active Learning: Cooperation in the College Classroom.* Edina, MN: Interaction Book Company.

- Johnson, D.W., Johnson, R. T, & Smith, K.A. (1991b). *Cooperative learning: increasing college faculty productivity.* Washington, D.C.: ASHE/ERIC Higher Education.

- McLeod, PL, SA Lobel, and TH Cox, Jr. 1996. *Ethnic diversity and creativity in small groups.* Small Group Research 27(2): 248-264.

- Michaelson, L.K. & Black, R.H. (1994). *Building learning teams: The key to harnessing the power of small groups in higher education.* In Collaborative Learning: A Sourcebook for Higher Education, Vol. 2, pp. 65-81. State College, PA: National Center for Teaching, Learning, and Assessment.

- Michaelson, L.K., Fink, L.D., & Knight, A. (1997). *Designing effective group activities: Lessons for classroom teaching and faculty development.* In D. Dezure (ed.) To Improve the Academy, Vol. 16, pp. 373-398. Stillwater, OK: POD Network.

- Oakley, B., Felder, R. M., Brent, R., & Elhajj, I. (2004). *Turning student groups into effective teams.* Journal of student centered learning, 2(1), 9-34.

[16] Think-Pair-Share" collaborative learning strategy
- Roediger III, H. L., & Karpicke, J. D. (2006). *Test-enhanced learning: Taking memory tests improves long-term retention.* Psychological science, 17(3), 249-255.

- Think (metacognition, reflection)

- Kaddoura, M. (2013). Think pair share: A teaching learning strategy to enhance students' critical thinking. Educational Research Quarterly, 36(4), 3-24.

- Bruffee, K. A. (1993). Collaborative learning: Higher education, interdependence, and the authority of knowledge. Baltimore, MD: Johns Hopkins University Press.

- Cabrera, A. F., Crissman, J. L., Bernal, E. M., Nora, A., Terenzini, P. T., & Pascarella, E. T. (2002). *Collaborative learning: Its impact on college students' development and diversity.* Journal of College Student Development, 43(1), 20-34.

- Davidson, N., & Major, C. H. (2014). *Boundary crossing: Cooperative learning, collaborative learning, and problem-based learning.* Journal on Excellence in College Teaching, 25 (3&4), 7-55.

- Dees, R. L. (1991). The role of cooperative leaning in increasing problem-solving ability in a college remedial course. Journal for Research in Mathematics Education, 22(5), 409-21.

- Gokhale, A. A. (1995). *Collaborative Learning enhances critical thinking.* Journal of Technology Education, 7(1).

[17] Benefits of making learning visible
- Hattie, J. (2015). The applicability of Visible Learning to higher education. Scholarship of teaching and Learning in Psychology, 1(1), 79.

- Hattie, J., & Yates, G. C. (2013). Visible learning and the science of how we learn. Routledge.

- Krechevsky, M., Mardell, B., Rivard, M., & Wilson, D. (2013). *Visible learners: Promoting Reggio-inspired approaches in all schools.* John Wiley & Sons.

- Ritchhart, R., Church, M., & Morrison, K. (2011). Making thinking visible: How to promote engagement, understanding, and independence for all learners. John Wiley & Sons.

[18] Doyle, Terry (2008). *Helping students learn in a learner centered environment: A guide to teaching in higher education.* Sterling, VA: Stylus.

[19] The importance of building community in online learning

- Garrison, D. R. (2016). E-learning in the 21st century: A community of inquiry framework for research and practice. Taylor & Francis.

- Nilson, L. B., & Goodson, L. A. (2017). Online teaching at its best: Merging instructional design with teaching and learning research. John Wiley & Sons.

- Boettcher, J. "*Ten best practices for teaching online.*" Quick Guide for New Online faculty (2011).